Tristan Tzschichholz

Passive satellite pose estimation based on PMD/CCD sensor data fusion

Tristan Tzschichholz

Passive satellite pose estimation based on PMD/CCD sensor data fusion

Südwestdeutscher Verlag für Hochschulschriften

Imprint
Any brand names and product names mentioned in this book are subject to trademark, brand or patent protection and are trademarks or registered trademarks of their respective holders. The use of brand names, product names, common names, trade names, product descriptions etc. even without a particular marking in this work is in no way to be construed to mean that such names may be regarded as unrestricted in respect of trademark and brand protection legislation and could thus be used by anyone.

Cover image: www.ingimage.com

Publisher:
Südwestdeutscher Verlag für Hochschulschriften
is a trademark of
International Book Market Service Ltd., member of OmniScriptum Publishing Group
17 Meldrum Street, Beau Bassin 71504, Mauritius

Printed at: see last page
ISBN: 978-3-8381-5010-9

Zugl. / Approved by: Würzburg. Universität, Diss., 2014

Copyright © Tristan Tzschichholz
Copyright © 2015 International Book Market Service Ltd., member of OmniScriptum Publishing Group
All rights reserved. Beau Bassin 2015

Abstract

For many problems, the estimation of the position and the orientation of an object is required, for example when regarding Rendezvous operations between satellites, what is of specific importance for future missions and the problem of space debris. In this regard, the PMD camera in conjunction with a CCD camera provides peculiar accuracy. However, for solving the problem, a few intermediate tasks are necessary.

For instance, the measurement range of the PMD camera is too small (7–10 m) and previous attempts of expanding it by estimating the phase shift across periods ("Phase unwrapping") always required a specific scene geometry, so that the phase shift could be recovered across periods by using actual phase jumps. This particular scene geometry just happens not to be present in space applications; typically, there is rather a single object visible, which freely floats in the image. Thus, in the thesis a new method for estimating the phase shift across periods is presented, which is suited for application in space. It is based on the beat principle, where a synthetic modulation frequency is generated by using two slightly different frequencies, which then allows measurements at larger distances. As a consequence, for the first time, the PMD camera becomes a viable option for Rendezvous maneuvers in space in this regard.

When an eligible distance image is available in conjunction with a grayscale image of the CCD camera, it must be decided whether the target object is visible, and if it is, where it is and how it is oriented. This task is usually being tackled by using template matching algorithms, but depending on the type of the sensor, other methods may be used as well. In particular, the principal component analysis (PCA) is a robust and well-established method for application on 3D point clouds. In the thesis, a method based on this is presented, which relies solely on the PMD camera, what brings along several advantages: complete independence of the position of the sun, quick image acquisition, large degree of detail. As a result, the position and the orientation of the target object can be estimated very reliably. Moreover, optionally, additional information in the form of different application-specific descriptions provide the ability of resolving ambiguities caused by object symmetry. Specifically, the robustness against changes in lighting and the combination with the PMD sensor is what makes this method superior.

As soon as a rough pose estimate (position and orientation) is available, it needs to be refined and updated over time, such that at each point in time relevant for the application, a preferably accurate pose estimate is available. Typically, laser scanners and stereo vision systems are used. However, stereo vision is limited in terms of distance and fairly sensitive with regard to environmental lighting. Laser scanners are robust, but comparatively expensive. Thus, in the thesis, the rather inexpensive fusion of a CCD camera and a PMD camera is proposed, where the fusion of image data is done by a reduction onto deviation components for the individual degrees of freedom of the pose estimate followed by a combination of the very same components. This new approach allows putting the complimentary advantages of the sensors to good use and ameliorates the impact of the drawbacks to a large degree, while reducing the required computation time to a minimum at the same time. Contrary to existing approaches of sensor data fusion using stochastic state estimators, such as the Kalman filter or the Particle filter, here, data that is known to be of low quality (e.g. data which has a partially predictable fuzziness) is purposely not included.

Finally, it is regarded as imperative to provide statements about the fitness and the accuracy of the method. To that end, an appropriately suited and highly accurately calibrated test environment – the European Proximity Operations Simulator (EPOS) of the German Aerospace Center (DLR) – could be used for evaluating the accuracy on one side, and for simulating preferably realistic environmental light on the other side. For the first time, a PMD camera could be used to measure precisely against a reference over larger distances ($>10\,\mathrm{m}$) on this facility. With a remaining error of about 6 cm on the distance, the method is approximately even with existing methods, but happens to be a lot less expensive and considerably less susceptible to mechanical stress than laser scanners.

A mistake in the firmware of the PMD camera was discovered during the measurements and later corrected by the manufacturer upon inquiry. This also posed an advanced test for the suppression of environmental light of the PMD camera which led to the conclusion that for a real deployment in space, hardware modifications are necessary – primarily, the use of a suitably defocused and modulated laser as light source and a preferably narrow optical band pass filter optimized for the wavelength of the laser.

Kurzfassung

Bei vielen Problemstellungen wird die Schätzung der Position und Orientierung eines Objektes benötigt, so zum Beispiel auch bei Rendezvous-Manövern zwischen Satelliten, was insbesondere mit Hinblick auf zukünftige Missionen und dem Problem des Weltraumschrotts von erhöhter Bedeutung ist. Die PMD-Kamera liefert hier zusammen mit einer CCD-Kamera eine besonders hohe Genauigkeit. Um die Aufgabe zu lösen, sind allerdings einige Zwischenschritte notwendig.

So ist z.B. der Distanzmessbereich der PMD-Kamera zu klein (7–10 m) und bisherige Versuche, diesen zu erweitern durch Schätzung der periodenübergreifenden Phasenverschiebung („Phase unwrapping") benötigen stets eine bestimmte Geometrie der Szene, um anhand von konkreten Phasensprüngen („Phase jumps") die periodenübergreifende Phasenverschiebung zu bestimmen. Diese Geometrie der Szene ist aber gerade im Weltraumbereich nicht vorhanden; viel mehr ist für gewöhnlich nur ein Objekt zu sehen, welches frei im Bild schwebt. In der Arbeit wird daher eine neue Methode zur Bestimmung der periodenübergreifenden Phasenverschiebung gezeigt, welche für die Bedingungen im Weltraum geeignet ist. Sie basiert auf dem Prinzip der Schwebung, wo durch Verwendung zweier leicht unterschiedlicher Frequenzen eine synthetische Modulationsfrequenz erzeugt wird, welche Messungen mit größeren Distanzen ermöglicht. Damit wird die PMD Kamera in dieser Hinsicht erstmalig sinnvoll für Rendezvousmanöver im Weltraum einsetzbar.

Liegt ein geeignetes Distanzbild zusammen mit einem Graustufenbild der CCD Kamera vor, muss zunächst entschieden werden, ob das Zielobjekt sichtbar ist, und wenn ja, wo es sich befindet und wie es orientiert ist. Diese Aufgabe wurde bislang oftmals mit Template-Matching-Algorithmen angegangen, abhängig von dem verwendeten Sensortyp kommen aber auch andere Verfahren zum Einsatz. Insbesondere hat sich hier bei 3D-Punktewolken die Hauptkomponentenanalyse (PCA) als robustes Verfahren durchgesetzt. In der Arbeit wird ein darauf basierendes Verfahren vorgestellt, welches ausschließlich mit der PMD-Kamera arbeitet, womit eine

Reihe von Vorteilen einhergehen: komplette Unabhängigkeit vom Stand der Sonne, schnelle Akquisition des Bildes, hoher Detailgrad. Die Position und Orientierung des Zielobjektes kann dadurch mit hoher Zuverlässigkeit geschätzt werden. Optionale Zusatzinformationen in Form verschiedener anwendungsbezogener Beschreibungen ermöglichen darüberhinaus die Auflösung der von der Objektsymmetrie verursachten Mehrdeutigkeiten. Insbesondere die Robustheit gegenüber Änderungen der Lichtverhältnisse macht dieses Verfahren in Kombination mit dem PMD Sensor überlegen.

Ist die ungefähre Lage (Position und Orientierung) des Zielobjektes bekannt, muss sie präzisiert und nachgeführt werden, sodass zu jedem für die Anwendung relevanten Zeitpunkt eine möglichst präzise Lageschätzung zur Verfügung steht. Typischerweise kommen hier Laserscanner oder Stereo-Vision-Systeme zum Einsatz. Stereo Vision ist allerdings limitiert in Bezug auf die Distanz und relativ empfindlich, was das umgebende Licht angeht. Laserscanner sind robust, aber vergleichsweise teuer. In der Arbeit wird daher die relativ kostengünstige Fusion einer CCD- und PMD-Kamera angeregt, wobei die Fusion der Bilddaten in Form einer Reduktion auf Abweichungsindikatoren für die einzelnen Freiheitsgrade der Lage mit anschließender Kombination derselben erfolgt. Dieser neue Ansatz ermöglicht es, die sich sehr gut ergänzenden Vorteile der beiden Sensoren optimal zu nutzen, ihre Nachteile weitgehend auszuräumen und gleichzeitig die benötigte Rechenzeit auf ein Minimum zu senken. Im Gegensatz zu bestehenden Ansätzen zur Sensordatenfusion mittels stochastischer Zustandsschätzer wie dem Kalman-Filter oder dem Particle-Filter werden hierbei Daten, über die von vornherein bekannt ist, dass sie von schlechter Qualität sind bzw. die sich durch eine partiell vorhersagbare Unschärfe auszeichnen, bewusst außen vor gelassen.

Abschließend gilt es, Aussagen über die Tauglichkeit und auch die Genauigkeit des Verfahrens zu machen. Zu diesem Zweck konnte eine entsprechend geeignete und hochgenau kalibrierte Testumgebung – der European Proximity Operations Simulator (EPOS) des Deutschen Zentrums für Luft- und Raumfahrt (DLR) – genutzt werden, um dort einerseits die Genauigkeit zu evaluieren und andererseits möglichst realistische Lichtverhältnisse zu simulieren. Erstmals konnte auf dieser Anlage mit einer PMD Kamera mit größeren Distanzen (>10 m) präzise gegen eine Referenz gemessen werden. Mit einem Restfehler von etwa 6 cm in der Distanz liegt das Verfahren etwa gleichauf mit bestehenden Methoden, ist dabei aber gleichzeitig erheblich kostengünstiger und mechanisch weitaus weniger empfindlich als Laserscanner.

Während der Messungen konnte ein Fehler in der Firmware der PMD Kamera entdeckt und vom Hersteller auf Anfrage behoben werden. Dies stellte auch einen weitergehenden Test der Fremdlichtunterdrückung der PMD-Kamera dar, der zum Ergebnis führte, dass für einen realen Weltraumeinsatz Hardwareänderungen an der PMD Kamera notwendig sind – vornehmlich der Einsatz eines geeignet aufgeweiteten modulierten Lasers als Lichtquelle und ein auf die entsprechende Wellenlänge optimierter, möglichst schmalbandiger optischer Bandpassfilter.

Acknowledgements

This work is the outcome of a cooperation between the University of Würzburg and the Spaceflight technology department (RB-RFT) of the German Aerospace Center (DLR). At this point, I would like to express my sincere gratitude to Prof. Dr. Klaus Schilling for his guidance and insight into scientific matters. I am deeply grateful for having had the opportunity to pursue this subject within this cooperative relationship.

I would also like to thank the head of the DLR spaceflight technology department, Dipl.-Math. Thomas Rupp, for making it possible for me to work at the DLR. It was a privilege being allowed to spend some time at an internationally well-known research institution and having a facility like the European Proximity Operations Simulator at my disposal. Apart from that, I was involved in a series of interesting projects which I would have otherwise not seen from the inside. It is the opportunities to learn and meet other people, which made a lasting impression on me.

This brings me to acknowledge the hard work and fruitful discussions with Dr.-Ing. Toralf Boge, head of the On-Orbit Servicing group. He was the technical contact person in all EPOS and research matters during my time at DLR. Thanks to his knowledge and experience, I could avoid encountering several pitfalls and streamline the process of making progress.

Thanks go also to my colleagues, Heike Benninghoff, for several related discussions and building a navigation filter using some of my signal processing algorithms and thereby putting the work into practical application. I would like to thank Tilman Wimmer, who made the first days and weeks at DLR easier for me. I would also like to thank Florian Rems for a lot of discussions and questions which helped me staying on track and focusing on the right problems. Also, I will never forget his unique way of making fun of problems. Furthermore, I am thankful for the hardware-related exchange with Andreas Grillenberger, as well as the provision of a digital storage oscilloscope for measuring the modulation signal of the PMD camera. The work of

Leonardo Regoli is also much appreciated, as he wrote the first prototype for the GigEVision camera driver. I am further grateful to David Macintosh for designing a mounting adapter for the PMD camera.

Thanks go also to Bernhard Thaler and the company Robo Technology GmbH for providing me with CAD drawings of several EPOS parts, which were relevant for this thesis. I would like to thank Markus Grothof and Martin Profittlich at PMDtec GmbH for their support and the updated firmware for the PMD camera, which was required in order to get a specific exposure mode up and running.

Finally, I would like to thank my family and friends for always believing in me and encouraging me. This would not have been possible without your support.

Contents

I	**Introduction**	**17**
1	**Overview**	**19**
1.1	Motivation	19
	1.1.1 On-Orbit Servicing	20
	1.1.2 Formation flying	22
1.2	Related work	23
1.3	Contributions	24
1.4	How this work is organized	26
	1.4.1 Overview flow diagrams	26
	1.4.2 Document outline	27
1.5	Math Notation	28
1.6	General conventions	30
1.7	Method summary	31
2	**Measurement, sensor and image processing fundamentals**	**35**
2.1	Introducing the EPOS facility	35
	2.1.1 History	36
	2.1.2 The new (and current) EPOS	38
	2.1.3 Other Hardware-In-The-Loop simulation facilities	41
2.2	Imaging principle	41
	2.2.1 Pinhole camera model	42
	2.2.2 Normalized coordinates	42
	2.2.3 Image space	43
2.3	CCD sensor	43
	2.3.1 Resolution and color	44

	2.3.2	Interface	44
	2.3.3	Optics	45
	2.3.4	Mechanical and power	46
	2.3.5	Final decision	46
	2.3.6	Sensor calibration	46
2.4	PMD sensor		49
	2.4.1	Theory of operation	51
	2.4.2	Consistency checking	58
	2.4.3	Sensor calibration	59
2.5	Extrinsic sensor calibration		60
	2.5.1	Initial estimation	61
	2.5.2	Angular refinement	64
2.6	Image processing methods		66
	2.6.1	Texture segmentation	66
2.7	Summary		69

II Algorithm design   71

3 Off-line preprocessing & Sensor characteristics   73

3.1	Sensor characteristics		73
	3.1.1	Lateral resolution	74
	3.1.2	Depth perception	74
	3.1.3	Frame rate	75
	3.1.4	Noise	75
	3.1.5	Separation of the search space	76
3.2	Model preprocessing		78
	3.2.1	Surface extraction	79
	3.2.2	Edge finding	80
	3.2.3	Surface centroids	85
	3.2.4	Principal components of surfaces	86
	3.2.5	Surface bounding box estimation	88
3.3	Direction vector preprocessing		88
3.4	Summary		89

4 PMD Data preprocessing — 91
4.1 Measurement error compensation — 91
4.1.1 Fixed pattern noise compensation — 94
4.1.2 Wiggling effect compensation — 96
4.1.3 Integration time offset compensation — 100
4.1.4 Amplitude offset compensation — 102
4.2 Range extension — 105
4.2.1 Multi-frequency phase unwrapping — 106
4.2.2 Noise considerations — 110
4.2.3 Noise/ range trade-off — 112
4.3 Summary — 114

5 Target acquisition — 115
5.1 Related work — 115
5.2 Initial pose estimation — 117
5.2.1 Architectural background — 117
5.2.2 Image-space coarse acquisition — 118
5.2.3 Dominant surface estimation — 120
5.2.4 Surface orientation — 122
5.2.5 Object symmetry related pose ambiguity — 127
5.3 Summary — 128

6 High accuracy fused target tracking — 129
6.1 Related work — 129
6.1.1 Scene acquisition — 130
6.1.2 Scene segmentation and object recognition — 131
6.1.3 Pose estimation — 132
6.2 Geometric feature estimation — 134
6.2.1 Dominant surface — 134
6.2.2 Surface affinity matrix — 134
6.3 Scanline processing — 137
6.3.1 Scanline parameters — 137
6.3.2 Texture change point — 139
6.3.3 2D image space deviation components — 141

    6.4   Point cloud processing 143
           6.4.1   Surface affinity determination 143
           6.4.2   Surface affinity matrix refinement 144
           6.4.3   Deviation component determination 144
    6.5   Pose estimation 145
           6.5.1   Pose representation 146
           6.5.2   Pose vector optimization 146
           6.5.3   Error metric 147
    6.6   Summary 148

III Measurements, Analysis & Discussion   149

7 Performance measurements   151
    7.1   Setup 151
           7.1.1   Sensor mounting 152
           7.1.2   Target mockup 152
           7.1.3   Sun simulation 153
           7.1.4   System structure 156
           7.1.5   Sensors 156
           7.1.6   PMD Sensor issues 157
    7.2   Reference measurements 161
           7.2.1   Linear approach 162
           7.2.2   Compound motion 168
    7.3   Static target performance evaluation 172
    7.4   Start-up process evaluation 173
           7.4.1   Initializer performance 173
           7.4.2   Tracker convergence 178
    7.5   Rendezvous scenarios 182
           7.5.1   Backside sun illumination 183
           7.5.2   Side sun illumination 185
           7.5.3   Frontal sun illumination 186
    7.6   Summary 189

8 Summary & Discussion — 191
8.1 Comparison of pose estimation methods — 191
8.1.1 PMD sensors — 191
8.1.2 CCD sensors — 193
8.1.3 Laser scanners — 194
8.1.4 Stereo vision — 194
8.1.5 Summary and overall performance comparison — 195
8.1.6 Kinect sensor — 197
8.2 PMD sensor space suitability considerations — 199
8.2.1 Thermal — 199
8.2.2 Radiation — 200
8.2.3 Launch vibration — 200
8.3 Conclusions — 201
8.3.1 Summary — 201
8.3.2 Algorithm limits — 203
8.4 Future work — 204
8.4.1 Feature-based docking tracker — 204
8.4.2 Extended range — 204
8.4.3 Hidden object removal — 205
8.4.4 Laser illumination — 206
8.4.5 Initializer object symmetry resolution — 206
8.4.6 Environmental effects — 207
8.5 Outlook — 207

List of Figures — 209

List of Tables — 213

List of Algorithms — 215

Nomenclature — 217

Bibliography — 223

Index — 243

Part I

Introduction

1 Overview

In this work, a method for estimating the position and orientation of uncooperative objects is presented. Uncooperative in this context means passive, i.e. the target object can not aid in a high-level process such as rendezvous or docking. It does not have actuators (or its actuators are disabled for some reason) and it does not have visual aids such as markers or reflectors. This particular task of visual object detection and tracking is encountered in several different application fields, of which the most prominent ones will now be briefly summarized.

1.1 Motivation

To date, there are thousands of man-made parts orbiting the earth. According to recent analyses, the number of parts above a size of $10 \times 10 \times 10$ cm is at about 19,000 [50] and still increasing. The objects originate from very different sources – mostly unmanned missions, but also manned missions. The exact number is difficult to tell, because there is a large number of objects for which information is not available. Figure 1.1 gives an impression of the space surrounding earth.

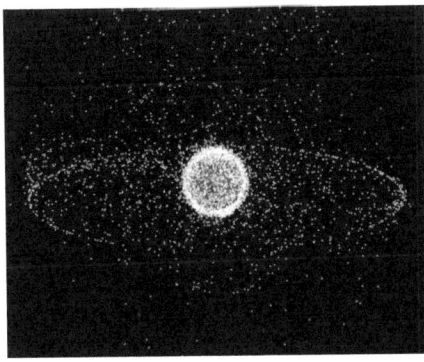

Figure 1.1. Debris in the space near to earth: In this image, some of the known debris parts are shown. The density is increased at low earth orbit (LEO) due to numerous missions targeting the area. Another accumulation can be seen as a circle (geostationary orbits). *(Image from NASA, public domain)*

What can be seen in the figure is that unlike from what one would probably expect, earth is surrounded by a significant amount of space debris. Space debris originates from collisions, deactivated or uncontrolled satellites, as well as rocket parts and even lost tools of astronauts. Most of these objects are detected and tracked by optical and radar measurements. Radar measurements work almost independently of the weather, but due to atmospheric dampening effects and power/ wavelength limitations, small objects can not be measured very well or not at all. Optical measurements with lasers require a clear sky. For more details on space debris measurements see [128].

Space debris poses a large risk to satellites and even to the International Space Station (ISS). This is a consequence of the large relative velocities of debris parts, which can reach several kilometers per second [7].

Interestingly, according to [50], an analysis performed by NASA would indicate that even if all nations contributing to space debris stopped launching more missions, the number of debris parts in low earth orbit (LEO, altitude range 80–2,000 km) would continue to increase for about 50 years as a result of collisions.

As a consequence, the need for space debris mitigation arose and On-Orbit Servicing became a prominent answer to the problem.

1.1.1 On-Orbit Servicing

No matter how many satellites there are, all of them have one problem in common: their lifespan. Depending on the mission, the average lifetime of a satellite is in the order of a few years. Large, expensive geostationary satellites last about 10–15 years. The lifetime has increased over the past decades [91], but is still considered being a relevant limit. Low earth orbit satellites have significantly shorter life cycles.

When looking at the statistics of reasons for satellites being rendered inoperable, one of the problems is the limited fuel supply. The fuel is needed to power thrusters, which are in turn used to keep the satellite in its orbit. Once the fuel runs out, the satellite is decommissioned.

This is why a typical application of On-Orbit Servicing is taking over attitude control. In this scenario, the servicer satellite will dock to the client satellite and use its own systems to stabilize the attitude of the now physically combined satellites. For the docking, for example in the OLEV mission, a specific docking tool was developed which can dock to the apogee thruster of a satellite [66]. This part is common to most satellites and as such, a good (if not the only point) where docking can be performed safely. Another approach is using a robotic manipulator, as it is planned for the DEOS mission [131].

What is also of interest is the ability to exchange or upgrade and repair single instruments of a satellite. This is especially of interest for geostationary satellites, which have been in orbit for a long time and for which the ongoing presence is still profitable [91, 122].

Finally, even if the satellite can not be used any longer, because a vital system has failed (according to [77], the most likely cause is power system failure), docking can make sense when the satellite poses a threat to other satellites in form of a collision target. Then, the satellite may be dragged into a graveyard orbit, or it may be safely deorbited. Graveyard orbits are locations where old satellites can reside without causing a risk to operational satellites or manned spaceflight missions. Deorbiting is the process of decreasing the altitude of a satellite, until it is also no longer a threat. Deorbited satellites are passive and have a quickly decaying orbit, what means they enter the atmosphere after a certain time and burn up.

A Rendezvous and Docking maneuver, what is required for On-Orbit Servicing, consists of several phases, of which roughly (1) is to synchronize the orbits of the two spacecraft, (2) is to decrease the relative distance between the active spacecraft (chaser) and the passive spacecraft (target) and (3) performing the actual docking process.

In practice, these phases are divided into even more subphases, where for each spacecraft, necessary conditions and requirements are specified along the goals [147].

It is now important to see that the process of docking requires a lot of on-board autonomy due to the latency involved in any form of communication with the satellite. This, in turn, requires accurate measurements of the relative position of the target satellite, depending on the way it is "seen" by the chaser. This particular problem of estimating the so-called relative *pose* (position and orientation) of a satellite, what is of utmost importance for such maneuvers, is investigated within this thesis.

There are a lot more applications for pose estimation. However, the work at hand will be limited to the space context, because the environmental conditions have a large impact on the design of the methods. Consequently, what works well in space does not necessarily (and most likely, will not) work well on earth and vice versa.

1.1.2 Formation flying

Another application field for visual navigation and pose estimation is formation flying of satellites. This is often mentioned together with applications where a satellite cluster can obtain performance/ cost ratios which would otherwise not be in reach [23]. Also, satellite construction is possibly standing before a paradigm shift towards networked, smaller satellites [126]. The relative distance and orientation of the satellites in a formation must be controlled with a high degree of accuracy. In this context, vision-based sensors are very likely to be used [146] among GPS receivers, depending on the mission and the design of the related satellites.

In general, sensors such as cameras and LiDAR sensors are often involved in the process of pose estimation [37, 101]. While to date, most of these systems require markers or some other kind of aid on target objects, the future draws a clearly different picture: The capability of autonomous object acquisition and tracking, combined with guidance systems, is required. This work focuses on the CCD and the PMD sensors (especially because the PMD sensor is a recent development with only little history and has yet never been used in a space environment) and an approach which does not require any aids present on the target.

Moreover, minimizing energy consumption, mass and complexity are goals which can be reached with these sensors. This has a large need for research as a consequence. For the very same reason, the work at hand is a small contribution to a large field, which requires continuous attention and development.

After having had the opportunity to contribute to an experiment performed on a pair of formation flying satellites (PRISMA) [103], the work can be put into context from a data processing point of view. Some of the challenges encountered during this experiment influenced the design of the method proposed in this thesis, despite the fact that the so-called ARGON experiment was targeted at far-range navigation only [28]. For example, the image resolution and -rate are similar. Also, it became clear that a monocular camera is sufficient for the task, as color information does not necessarily provide a valuable input for pose estimation (as will be seen in the literature review, RGB cameras are almost never used in space contexts). Besides, color triples the memory requirements for stored frames and in-memory processing. More details about the selection of the sensors can be found in section 2.3 on page 43 (for the CCD camera) and in section 2.4 on page 49 (for the PMD camera).

1.2 Related work

Autonomous Rendezvous and Docking is receiving persistent attention from the research community, as well as companies designing space-related components and systems. Various theoretical works and some practically usable sensor systems have emerged, of which a few prominent ones will be briefly reviewed in the following.

With TriDAR [37], an interesting solution of the pose estimation problem is available, combining the LiDAR approach (time of flight) with triangulation (have two emitters point at a single spot on the object, then get the distance from the bearing angles). The idea behind the approach is to combine the advantages of both methods: The long range capabilities of a LiDAR sensor and the accuracy of a triangulation approach in the near range. The sensor is not only suitable for Rendezvous and Docking problems, but also for planetary landing and even rover navigation. Unfortunately, the resulting sensor system is expensive when compared to purely camera-based systems.

A more inexpensive method is to rely on monocular vision only, but extended by a high-level reasoning layer [107]. This subsystem would evaluate the raw data of the image processing part and carry out assigned tasks. In the event of tracking failure or any other unforeseen problem with estimating the position or orientation of the target object, the subsystem would suggest corrections (retry, retry after random time interval) in order to solve the problem. This specific part is called "hybrid deliberative/ reactive computational framework". It can be summarized as a sophisticated mixture of several artificial intelligence disciplines used to enhance the overall performance of the system.

In contrast, rather common is the use of a scanning LiDAR sensor for estimating the pose of the target [63]. The target is detected in a first step and bearing and range is estimated. Then, a 3D model is fitted, after which the full pose is available. For following the target and updating the pose, a tracker is proposed. The correct knowledge of the pose then allows more efficient flight trajectories because the orientation and the position of the docking port is known at all times, even when it is not visible. The paper demonstrates the system as a proof-of-concept using a scaled satellite in a testbed.

The successor of the sensor used in the Orbital Express mission- the Automated Video Guidance Sensor (AVGS)- is primarily intended for the crew exploration vehicle (CEV) [80]. Two different lasers are used to illuminate the target. The difference image yields the remaining reflections from the reflectors with all other error sources removed. After a thresholding step, centroids of the reflections are determined and then the pose of the object is estimated. This is a high-cost (two lasers) solution, which also requires special reflectors at the target.

To summarize, a lot of projects related to autonomous Rendezvous and Docking use laser-based sensors. Often, fiducial markers must be present on the target. There are a few exceptions, but these are very rare and the resulting systems are very expensive (for example, the TriDAR system). Monocular vision is rarely used for full 6-DoF pose estimation, apparently due to accuracy concerns. Stereo vision is more common, but still seldom proposed in this context. LiDAR-based sensors are clearly dominating in this area.

Apparently, the possibility of directly and reliably measuring the distance is of utmost importance, therefore the method proposed in this thesis looks promising after a review of the relevant literature and may especially present an alternative to close-range laser-based sensors or purely CCD camera-based approaches.

The review of related work is continued throughout the work in form of trailing subsections regarding different aspects of the implementation. It was chosen to structure the work at hand in this way, as it touches a large number of different scientific fields.

1.3 Contributions

This work aims to implement a relative pose estimation method using a PMD camera and a CCD camera. While the CCD camera has been widely used in space already, the PMD camera has never been used in a space environment. This is why an investigation of the sensor with regard to space applications is particularly interesting. The work at hand provides:

- **A method to extend the measurement range of PMD sensors.** Because common PMD sensors are limited to relatively short distance ranges due to the measurement principle, a method is presented and evaluated which allows measuring at significantly larger distances; in the proposed scenario, the theoretical gain of measurement range is at 1,000 %. In contrast to existing methods, this approach does not rely on the scene geometry.

  A complete calibration framework based on available literature about PMD sensor calibration is given, which addresses the usage of two measurements from the beginning in order to reduce side effects in the fused long-range image. As state-of-the-art PMD camera calibration do not consider long-range measurements, known models are extended to accomodate the problems encountered in these situations.

  During the design and test phase, firmware problems have been found and identified in cooperation with the PMD sensor manufacturer, which are likely to have never been discovered before, because PMD sensors have not yet been used on calibrated test facilities as large as EPOS.

- **A fast, initializing pose estimator.** Most high-accuracy, real-time pose estimation problems can only be solved by designing a tracker architecture, which refines a given pose in a very short amount of time and still provides high-accuracy measurements. However, without finding an initial pose, the algorithm can not be used autonomously.

  In this work, such an initial pose estimation method capable of finding the target object in the image without prior information, is presented. The method draws its robustness from apparent benefits of the PMD sensor.

  Furthermore, the problem of object symmetry is ameliorated by adding an interface to the initializer what allows providing *hints*, i.e., additional mission- or situation-specific information, making it possible for the initializer to determine the correct object pose even for symmetric objects.

  A novelty is the usage of the distinct advantages of the PMD camera in combination with powerful tools such as the principal component analyis in order to retrieve a computationally efficient and robust result.

- **A multi-sensor fused high-accuracy pose tracker.** Investigations have shown that the PMD sensor and the CCD sensor have a complimentary nature when it comes to their advantages and disadvantages. A method is presented, which allows combining the advantages and ameliorating the shortcomings by fusing the data of the sensors in an application-specific representation.

  This is different from most established methods, where sensor data fusion is performed on the image level without specific regard to the high-level problem.

- **An in-depth analysis of the PMD sensor for use in space environments.** Due to the much more problematic solar irradiance in space, measurements have been made on a calibrated Rendezvous and Docking test facility in order to evaluate the fitness of the PMD sensor for use in space from a lighting-related point of view.

  Measurements have shown that simulating extraterrestrial sunlight is a complex subject both in theory and practice. Using a high-power sun simulation including a measurement of the light spectrum, tests could be performed which are significantly closer to the real situation as it would be encountered in orbit, when compared to previous approaches.

  Furthermore, an investigation of the modulation signals of the PMD camera is provided, what gives an insight into potential problems. A brief analysis of when they are encountered and how to compensate for them supplements the findings.

1.4 How this work is organized

1.4.1 Overview flow diagrams

For the main part of this thesis (the description of the pose estimation method), there are two flow diagrams, which summarize the chapters and sections, with the intention of providing an overview to the reader in order to facilitate understanding the process. These flow diagrams can be found on pages 32 (for the preprocessing part) and 33 (for the principal part).

1.4.2 Document outline

It was decided to structure this thesis into several chapters, which are itself wrapped into three large parts. The three parts are:

- **Part 1.** An introductory part which is intended to familiarize the reader with the methods and sensors used. Also, the reasoning for choosing the particular sensors is given. There are two chapters:
    - **Chapter 1.** The first chapter contains an overview of the work, as well as a literature review. It provides reasoning and the background of the work at hand and is also supposed to open up a broader view of the problem.
    - **Chapter 2.** The second chapter is dedicated to the brief introduction of basic methods, the chosen sensors and the measurement equipment. The purpose is to familiarize with typical error sources and introduce compensation methods.
- **Part 2.** The second part represents the core of this work. Here, the details of the data processing are presented. Where appropriate, relevant literature is briefly reviewed as well. The following chapters belong to part 2:
    - **Chapter 3.** In this chapter, off-line preprocessing of data is described. This step has been introduced in order to lower the computational burden in the on-line phase of the algorithm. Also, the sensor characteristics are investigated, which will later have a significant impact on the design of the algorithm.
    - **Chapter 4.** This chapter is dedicated to PMD sensor data preprocessing. The purpose of the chapter is twofold. First of all, it is supposed to show that serious effort must be put into processing PMD data, before it can be safely used – especially when using two frequencies at the same time. Second, a range extension method is presented, which allows escaping from the inherent distance measurement limitation common to PMD sensors.
    - **Chapter 5.** This chapter is dedicated to the initialization problem. A method will be presented, which uses the PMD sensor to estimate a

rough pose estimate without any prior knowledge, except for the target object model.

– **Chapter 6.** In this chapter, the target tracking algorithm is presented in detail. The method refines a given pose estimate using both sensor measurements from the PMD sensor and the CCD sensor and fuses the data streams on a rather high, application-specific level.

- **Part 3.** The last part of the thesis is dedicated to presenting the results achieved so far, and moves on to a comparison with other methods and sensors. The purpose of this part is to reflect on the results and provide a context along with a meaning. The part is divided into the following two chapters:

    – **Chapter 7.** This chapter is intended for to presenting the experiments performed, both in terms of what exactly has been done and why it has been done this way. Furthermore, the results are presented along with an interpretation.

    – **Chapter 8.** The last chapter provides a more in-depth analysis of the results, along with comparisons to other implementations and sensors. The purpose is to provide a broader view of the pose estimation topic and a personal estimation of how the work at hand fits into the field.

1.5 Math Notation

The following notation is applied throughout the document. It complies to established literature to a large degree, but some exceptions may be encountered. On occasion, the text may define slightly different notation where it eases the handling of a specific problem. In these cases, a profound description and reasoning will be provided.

a	A scalar.
a	A vector.
ã	Intermediate vector; often used to emphasize the relationship with **a** in long calculations.

(continued on the next page)

(continued from the previous page)

a_k The k-th element of the vector **a**. If k is not defined further, alternative notation for a set of vector elements. Used sometimes to make equations more readable.

A   A matrix.

$\mathbf{A}^*$ The adjugate of the matrix **A**.

$\mathbf{A}^T$ The transpose of the matrix **A**.

$\mathbf{A}[i,j]$ A specific matrix element (row indicated by i, column indicated by j).

$\hat{x}$ A vector or matrix (depending on the appearance of the symbol under the hat) transformed using a contextual transform, for example Fourier transform.

$\text{Im}(x)$ The imaginary part of a complex scalar x. When applied to a vector or matrix, it is applied element-wise.

$\text{Re}(x)$ The real part of a complex scalar x. When applied to a vector or matrix, it is applied element-wise.

$[a,b)$ A half-open interval; a is included, b is excluded. Round brackets indicate exclusion of the interval boundary, square brackets indicate inclusion of the interval boundary.

$\|\mathbf{x}\|$ The norm of the vector **x**. When not specified, the Euclidean norm $\|\cdot\|_2$ is used.

$\|\mathbf{x}\|_p$ The p-norm of the vector **x**, i.e. $\|\mathbf{x}\|_p = \left(\sum_{i=1}^n \|x_i\|^p\right)^{1/p}$, where n is the number of elements in **x**.

$\mathcal{A}$ A named set.

$\|\mathcal{X}\|$ The number of elements in the set $\mathcal{X}$.

$\emptyset$ The empty set.

$\angle(\mathbf{x},\mathbf{y})$ Angle between vectors **x** and **y**, $\angle(\mathbf{x},\mathbf{y}) = \arccos\left(\frac{\mathbf{x}\cdot\mathbf{y}}{\|\mathbf{x}\|\cdot\|\mathbf{y}\|}\right)$

(continued on the next page)

(continued from the previous page)

$\bar{q}$	The inverse of the quaternion **q**. Sometimes, the line is used to indicate a particular state of processing a vector. In these cases, it is made clear in the text.
q^*	The (complex) conjugate of the quaternion **q**.
$\odot$	The quaternion multiplication.
q (x)	Application of the quaternion **q** to a vector **x**, i.e. **q** (**x**) = (v_1, v_2, v_3), where **v** = **q** $\odot$ (**x**, 0) $\odot$ $\bar{\mathbf{q}}$.

Quaternions are treated as defined in [148], with the vector components located at the first three components of the quaternion, and the scalar component being the last element of the quaternion.

1.6 General conventions

Throughout this document, PMD stands for "Photonic Mixer Device", a semiconductor chip design which allows phase-shift measurements in hardware using modulated light. In a lot of documents, this is also referred to as "Time of Flight" (ToF), what is considered being a too imprecise (if not misleading) term in the authors opinion, since the time of flight is neither measured with these sensors nor is it directly relevant. Instead, the phase shift of a continuously emitted signal is measured and compared to a reference signal. The abbreviation "PMD" is therefore used to stress this difference.

Data transfer speeds are typically provided using the base of 1,000 instead of the IEC base 1,024[1]. While the capital "B" means byte, the letter "b" stands for bit. For storage sizes, the base 1,024 is used and the corresponding magnitude abbreviations are used such as GiB, MiB, and so on.

Euler angles are provided in pitch (up/ down) – yaw (left/ right) – roll order. They are used to visualize angular deviations, as it is difficult to interpret compound rotations from quaternion components.

---
[1] See IEC 60027-2, Ed. 3.0, (2005-08) for details.

Images from sensors have pixel numbers written on the X- and Y-axis. In the relevant plots throughout the document, these axes are not described further, as this is often not necessary to understand the content of a figure and to save space, the axes are also left without annotation.

1.7 Method summary

In the following, the entire algorithm will be summarized in form of two flow diagrams in order to get a quick overview. The flow diagrams will also mention the relevant chapters and sections where details about particular parts of the algorithm can be found.

First of all, the envisaged method is divided into a principal processing part (figure 1.3) and a preprocessing part (figure 1.2), and the preprocessing part is divided into an online phase and an offline phase.

In the offline phase, sensor data independent processing takes place, such as obtaining optimized model data from a raw model and obtaining auxiliary data for working with the PMD sensor. In the online phase, relevant sensor data is processed and transformed into suitable representations used by the principal part of the algorithm.

In the principal part of the algorithm, the two data streams of the PMD sensor and the CCD sensor are used to determine so-called deviation components, that is, sensor- and image-space specific measures of deviations between the current estimated location and orientation of the target object and the object captured by the sensors. The deviations are specific for the six degrees of freedom, as will be seen later. By minimizing these deviations for each image retrieved from the sensors, the estimated position and orientation of the target object is updated and corrected over time.

The reasoning for designing the algorithm this way and all the details about the individual steps will be provided in the referenced sections and chapters.

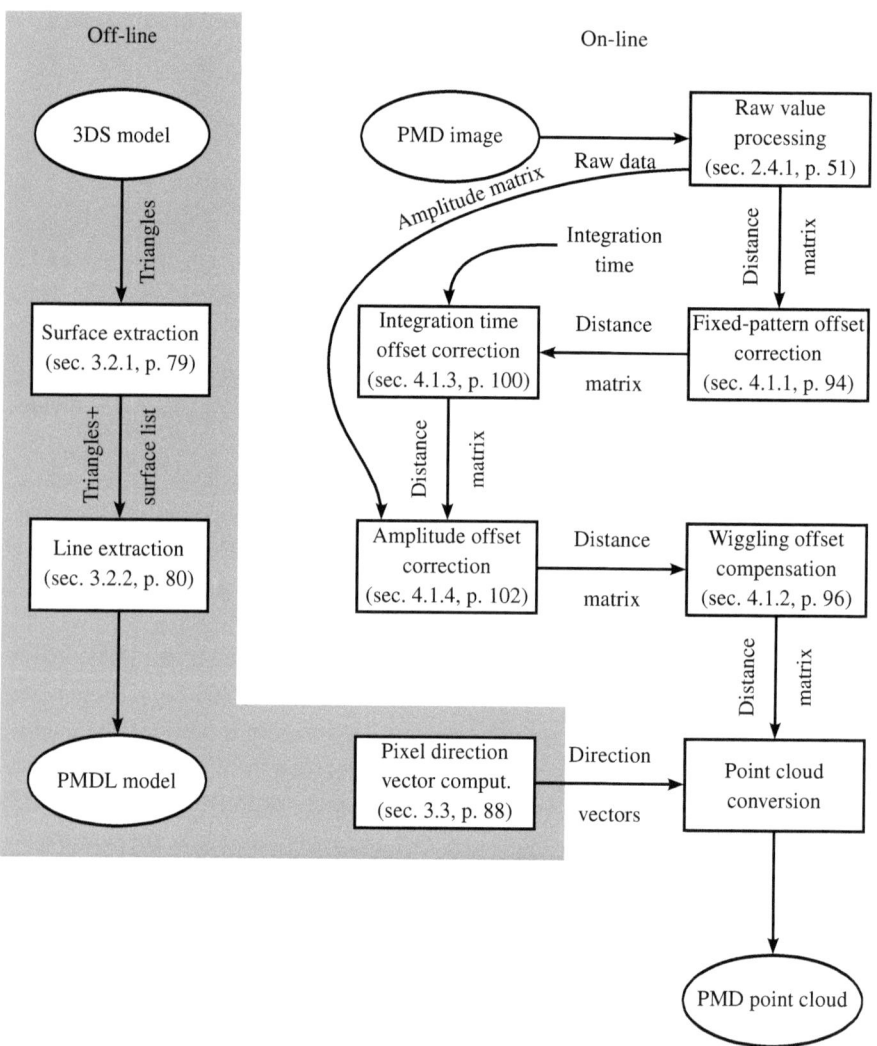

Figure 1.2. Algorithm flow chart (preprocessing part).

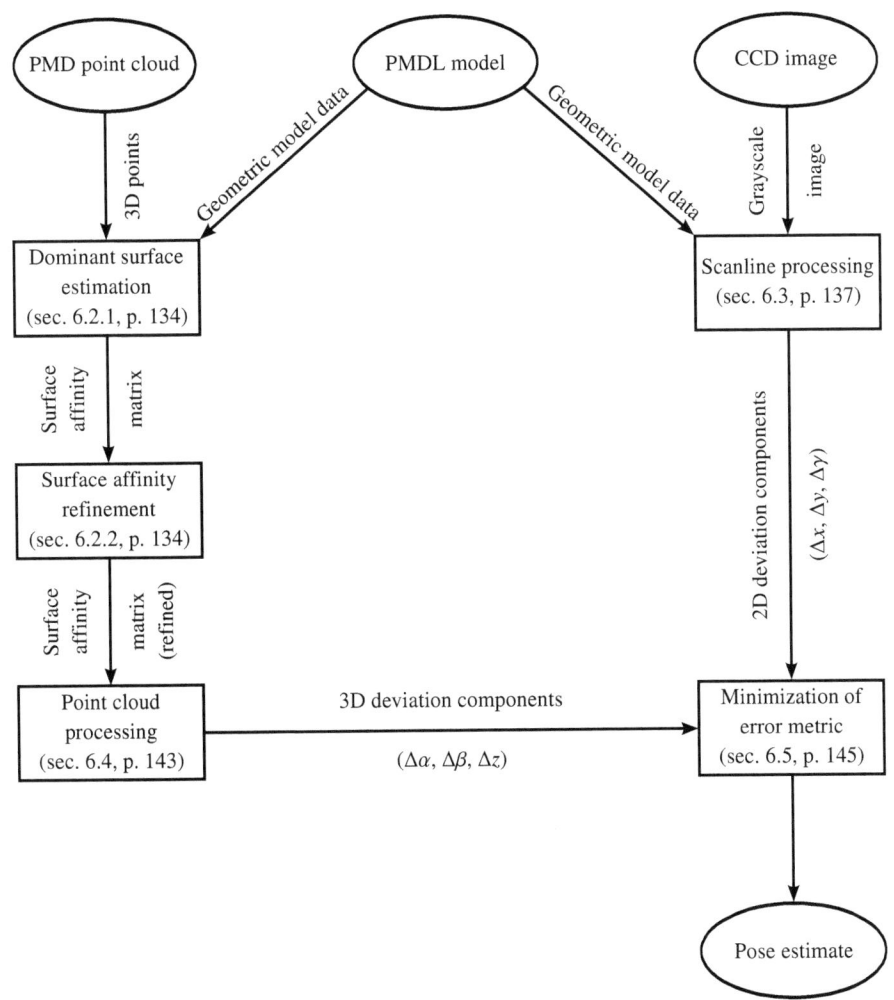

Figure 1.3. Algorithm flow chart (principal part).

2 Measurement, sensor and image processing fundamentals

The PMD sensor is a unique sensor in the sense that it requires specific signal processing applied to its output in order to retrieve a useable signal. At the same time, its abilities are special as well, as it combines a high frame rate with low power requirements and the parallelized distance measurements for every single pixel. Before starting to work with the sensor, it is therefore important to familiarize oneself with it. The same applies to the CCD sensor, although it is less substantial.

As it is the case with any pose estimation method, at the end, the question is *how accurate is it?* – To answer this question, measurements will be made on a facility specifically designed for testing Rendezvous and Docking payloads. To understand how the results are retrieved, it is essential to know the test environment. Therefore, this testing facility will be introduced as well.

Finally, the method presented within this thesis is built upon previous research work. Consequentially, it is important to have a brief look at these techniques first in order to understand how they are used to put together the entirety.

2.1 Introducing the EPOS facility

This section will briefly introduce the European Proximity Operations Simulator (EPOS), a real-time, hardware-in-the-loop Rendezvous and Docking simulation facility, which was used to simulate realistic docking trajectories and maneuvers in order to evaluate the performance of the image processing algorithms.

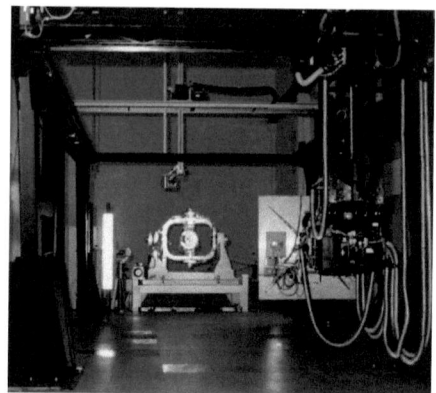

Figure 2.1. The old EPOS facility. A very large portal robot moves a gantry which provides a six degrees of freedom movement of the mounted sensors together with the robot. At the other end, a gantry was installed on ground which was capable of rotating about all three axes. *(Image courtesy of DLR)*

2.1.1 History

Origins

The EPOS facility was first built in cooperation of the European Space Agency (ESA) and the German Aerospace Center (DLR) and installed at the DLR site in Oberpfaffenhofen, Germany. At the same location, the German Space Operations Center can be found. The latter has been involved in countless projects ranging from satellite launches (launch and early orbit phase, LEOP) to satellite control, as well as several missions involving the International Space Station (ISS). For example, the Columbus module has its control center also in Oberpfaffenhofen.

The first version of the EPOS simulator has been started in 1984. The lab, which contains the simulator, has a size of about 40 × 10 m. It consisted of a very large portal robot which could move a gantry in X, Y and Z direction. This allowed to simulate a distance of about 11 meters with original size models. In case a larger distance is to be simulated, the model size can be reduced and the simulated distance would be increased at the same time by the corresponding factor.

The gantry was able to rotate the mockup about all three axes. This way, all six degrees of freedom are available. The mockup at the other end was also mounted on a gantry. This gantry was also capable of rotating the mockup about all three axes, what provided more flexibility.

The control system of EPOS was designed for real-time hardware-in-the-loop simulation, meaning the complete sensor and processing hardware of the system under test could be mounted on the EPOS actuators and tested on EPOS as a whole. While the sensors provide information about the position of the other mockup, the control system would produce thruster commands. The thrusters, however, are not implemented as real hardware, but simulated by EPOS. This requires that the dynamic models and orbit information is provided to EPOS.

As a result, navigation, orbit-keeping as well as Rendezvous and Docking systems could be tested in real-time. EPOS also featured a realistic illumination unit, which was capable of simulating the sun according to the simulated position of the two mockups in orbit. A theater lamp was used to provide the lighting. The main application EPOS was intended for was the simulation of Rendezvous and Docking maneuvers without contact dynamics.

The Phase A study was completed in 1985 which included the hardware design provided by DLR and the software by ESA. The cooperation partner was ESTEC in Sweden.

Phase B was completed in 1988, which defined the most important hardware and software concepts and provided the most important functional parts as prototypes. Thus, a firstly working simulation environment was complete.

Phase C1 (completed in 1995) then moved the prototype implementation into the building where the EPOS simulator is located until today. Furthermore, a lot of software development was done: DLR developed application software and kinematics, more simulation and test control software was provided by Origin and flight dynamics/ equipment modeling software was added by DASA. DASA was a company which belonged to Daimler Benz and could be seen as an aerospace division. It existed in original form until 2000, when it was changed to a pure holding society.

Phase C2 (completed in 1998) modernized several hardware components and brought along an interface to Eurosim. Eurosim is a configurable simulator tool for space-related real-time simulations. At the same time, the final preparations for the ATV simulations where made.

Modernisation

The very first version of EPOS was modernized in 2002, when the command frequency was raised from 2 Hz to 25 Hz [19]. Furthermore, the core computer was replaced by a more powerful one. The old VME bus (Versa Module Eurocard) sys-

Figure 2.2. The new EPOS facility. Two industry-grade manipulators simulate the motion of satellite mockups. A theater lamp is used to simulate the sun and a DLP projector can be used to provide an earth albedo or background. *(Image courtesy of DLR)*

tem[1] was replaced by a PC-based version, with the interconnections made with fiber cables. Remote access was implemented since the Internet had evolved rapidly.

Apart from that, several enhancements have been added. To mention a few, kinematics became exchangeable, a configuration management system was added and a sophisticated logging part was implemented.

The man-machine interface was improved as well. First of all, a preview capability was added, which provided the user with information on what would actually happen when a certain trajectory was executed. The visualization was in 3D, an important step forward in helping users to avoid crashes.

Figure 2.1 shows an image of this implementation. At the far end of the picture, one can see the stationary gantry. The large, black portal robot with the XYZ stage and the mobile gantry can be seen as well in the middle of the image.

Using the modernized version of EPOS, sensor tests of the ATV (Autonomous Transfer Vehicle) have been successfully performed.

2.1.2 The new (and current) EPOS

The facility was renewed in 2008, using two industry-grade KUKA manipulators, as shown in figure 2.2. The main drivers for this decision were increased accuracy, reduced maintenance overhead and reduced dependence on the manufacturer of rare parts. The control system is now completely PC-based with Windows PCs as a user interface and computers running VxWorks on the real-time side.

---

[1]The current version of this bus system, called VME64, is still in use in the aerospace industry. For example, several parts of the computer system of the International Space Station (ISS) are based on it.

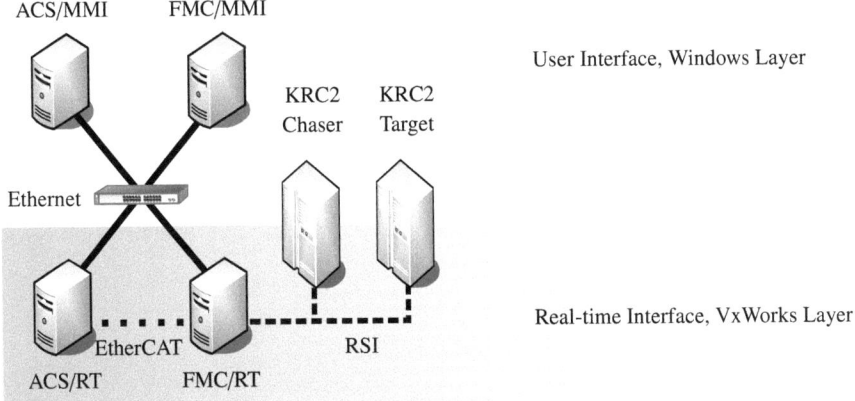

Figure 2.3. The architecture of EPOS. Two real-time machines running VxWorks command the robots and run the simulation, while Windows machines provide the user interface.

The architecture of the new EPOS facility employs the two KUKA manipulators as well as a 25 m rail. One of the robots is mounted on the rail. The two robot controllers are connected to the first real time machine, the so-called Facility Monitoring and Control (FMC-RT) (RT for real-time) via Ethernet. The FMC-RT then sends the movement commands to the robot. It is connected to a real-time capable bus system known as EtherCAT. EtherCAT connects the safety electronics and the Application Control System (ACS) where a user can run and design simulations on.

As an improvement, work is being concentrated on implementing contact dynamics for the simulator [153], so it can simulate the contact forces and actually simulate a real docking maneuver (including everything that can possibly go wrong). To that end, a force/torque sensor has been installed at the chaser mockup, between the docking tool and the flange.

To improve the accuracy of the robots, a special optical measurement system is being developed. It employs two units, of which one of both is mounted on each robot. Using a laser beam, the relative distance, displacement and orientation is measured and used to correct the positions of the robots relative to each other. As a result, when looking at relative positioning, the accuracy is pushed towards today's technical limits.

Table 2.1. Comparison of EPOS facility performance indices. The EPOS facility will further gain almost one magnitude of accuracy, once the optical measurement system is available. A special firmware supplied by KUKA allows commanding the robots at 250 Hz.

Property		pre-2008*	post-2008[†]	Unit
Accuracy	position	< 5.00	1.50	mm
	orientation	< 0.2	0.2	deg
Commanding rate		25	250	Hz
Maximum load	target	40	240	kg
	chaser	100	100	kg
Maximum velocity	translation	0.5	2	m s^{-1}
	rotation	6.0	180	deg s^{-1}
Approach distance range		11	25	m

* Including modernization of 2002. Values from [16, 150].
[†] Details can be found in [18, 119].

At both mockups, several ports are available in order to retrieve information from the sensors. The complete interface includes

- A 230 V line, for local power supplies, etc.

- Several DC power supplies including 5V/ 4A, 12V/ 2.5A, 24V/ 2.5A and 28V/ 2.5A.

- Two Ethernet ports.

- One RG-58 coaxial cable for RF transceivers/ GPS signal simulation.

- 18 × 1 mm² multi-purpose lines, unshielded.

The multi-purpose lines can be directly accessed in the EPOS control room what allows special equipment for sensor tests installed directly in there.

To conclude the evolution of EPOS, table 2.1 summarizes the most important technical differences between the generations.

2.1.3 Other Hardware-In-The-Loop simulation facilities

A similar facility can be found at the Marshall Space Flight Center in its Flight Robotics Laboratory (FRL) [118]. Here an air bearing floor is combined with a 8 degrees of freedom overhead gantry robot, which simulates the target object. Similar to EPOS, force/torque sensors are used to simulate contact dynamics. A sun/ lighting simulator is also available. A similar facility can be found in italy [53].

More similarities with EPOS are seen in the EPOSx facility near Rouen in France. Here, the same goals are set, but at an even larger scale. Distances of about 300 meters can be simulated, and the target weight can exceed 1,000 kg [24].

However, more common is the use of two fixed 6-DOF robots without a rail [63]. Here, two Fanuc robots accomplish the hardware side of the simulation.

The On-orbit Visual Environment Simulator is a scaled version of two robots, which allows simulation of Rendezvous and Docking maneuvers including Hardware-In-The-Loop simulation and sunlight for scaled (1:10) models [138].

To summarize, facilities like EPOS can be found all over the globe. However, facilities of the size of EPOS are rare and it is safe to assume, that there are not many that large.

2.2 Imaging principle

After having introduced the test facility, a brief introduction into optical sensors (and characteristic measurement errors) will be provided. As it is the case with all sensors, the measurement errors must be understood before calibration measures can be developed and applied. Fortunately, at least as far as optical distortions are concerned, the same method can be applied to both sensors. In this section, a model commonly found throughout the literature is introduced which facilitates the use of optical sensors tremendously.

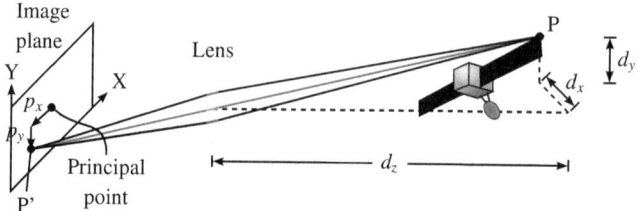

Figure 2.4. Perspective projection of a point P(d_x, d_y, d_z) and its corresponding image-space coordinates P'(p_x,p_y).

2.2.1 Pinhole camera model

Both sensors are modeled using the so-called *pinhole camera model* [57]. In this simplified model, a scene is captured by a camera with an infinitely small hole. For practical reasons, assuming that each point of the scene scatters light in deviating directions, the hole is replaced by a collimating lens, such that additionally scattered light is also seen on the same projection point (figure 2.4). Consequently, the principle remains the same when using a lens and the model as such does not need to be altered.

The only difference is that there is now optical distortion (which must be compensated) and points are no longer imaged as points but rather as circles (the image may become blurry), depending on the distance of the object and the focal length of the lens. By tuning the parameters correctly, a sharp image can be retrieved.

Despite the deviations of real-world image sensors from the original ideal model, this model extension is sufficient and commonly used in most image processing applications.

2.2.2 Normalized coordinates

An object is imaged by transforming all 3D points of the model into image space, i.e. 2D coordinates on the image plane, what corresponds to the surface of the sensor. What can be seen from figure 2.4 is that it is sufficient to capture the *direction* of a line of sight for each point of the object. This direction information is essentially encoded in *normalized coordinates*, because

$$\kappa = \left(d_x/d_z, d_y/d_z\right) = (\tan\alpha, \tan\beta), \tag{2.2.1}$$

where κ is the normalized coordinate vector and α and β are the direction angles (azimuth and elevation). Once the normalized coordinates are known, the image-space coordinates can be calculated by multiplying with the focal length (which is equal to the distance between the sensor and the lens), what yields

$$\mathbf{p} = \kappa f. \qquad (2.2.2)$$

Because $\mathbf{p}$ is relative to the central point of the image (which will also be referred to as *principal point*), by adding the pixel offset $\mathbf{c}$ of the latter, pixel indices $(i, j) = \mathbf{p}+\mathbf{c}$ are retrieved. Normalized coordinates are specifically suited for the correction of distortions which is why distortion will always be applied to or compensated on normalized coordinates. This will be discussed in more detail in sections 2.3.6 (page 46) and 2.4.3 (page 59).

2.2.3 Image space

At this point, a clarification about the term "image space" will be made. Throughout this work, this term refers to a two dimensional vector space of image coordinates.

Definition 2.1 (Image space) *The image space of a sensor is a two-dimensional vector space spanned by two sensor-specific intervals*

$$\Omega := ([0, w-1] \times [0, h-1]) \subset \mathbb{R}^2,$$

where $w \in \mathbb{N}$ is the width of the image and $h \in \mathbb{N}$ is the height of the image (in pixels).

Note that this definition allows locating points between pixels ("subpixels"). Also, the size of the image space depends on the sensor. In the text, this may be indicated by subscripts such as Ω_{PMD} or Ω_{CCD}, respectively.

2.3 CCD sensor

The monocular camera was first used without the PMD camera for initial tests and also for developing a pose estimation method, which would rely on a single sensor only for later comparison. The results can be found in [141, 142]. There are a lot of sensors available with varying suitability and targeted application. In the following, the reasoning for choosing this particular camera is given.

2.3.1 Resolution and color

On the ATV, a monochrome sensor (resolution 1,024×1,024 pixels [75]) based on a star tracker [145] is used in the docking assembly. The thought of a higher resolution is – and will be even more in the future – a promising one. However, the images generated by the sensor need to be processed in some way and, generally speaking, the higher the resolution of the image, the higher the processing time. Therefore, it seemed adequate to use a sensor with a moderate image resolution. Without investigating the problem in more detail, it seemed to be a reasonable choice of using a monochrome camera with VGA (640 × 480 pixels) resolution.

2.3.2 Interface

With two parameters set, a lot of other parameters remained open. As the intention is to use the camera together in a closed loop configuration, it is essential that the sensor can be used in conjunction with a real-time system. Basically four interfaces are available:

- **CameraLink**: A high-speed interface for digital video cameras used in industry automation. CameraLink comes in several speed grades and always requires dedicated hardware (e.g., frame grabbers). The maximum cable length depends on the speed grade and is at best 10 meters.

- **Firewire (IEEE 1394a/b)**: A serial bus system which can be seen as the de-facto bus standard for digital video connectivity. While Firewire can be commonly found on most computers and consumer products, it is not very common in research and even less in industrial applications.

- **Ethernet (IEEE 802.3)/GigEVision**: A very young standard which is built upon Ethernet. Ethernet currently goes up to 10 Gb/s, which is more than enough to transport even uncompressed image data. Most cameras come, however, in the lower speed grade (1 Gb/s).

- **Universal Serial Bus (USB)**: Found on all PCs and even servers, embedded and industrial products, the universal serial bus has a large footprint in device connectivity. The drawback is that each device requires a device-specific driver – when not open sourced, for each specific architecture and operating system.[3]

When it comes to the interface, one important constraint ruled out most of the interfaces at once: cable length. While Ethernet can operate on cable lengths up to 100 meters, all other standards allow for a maximum of 10 meters. Some of them can be extended using expensive repeaters. However, in an environment such as the EPOS facility, repeaters are not an option as any construction using repeaters poses a significant problem to the 25 meter rail system with underground cabling. Media converters were not available at the time the measurements have been made or have been considered too expensive.

Another problem is the availability of drivers for the real-time system. While Ethernet was up and running already, USB and Firewire were not available and for CameraLink, there would not have even been a driver available for a single frame grabber board.

In conclusion, it was decided to buy a camera with an Ethernet interface and write a custom driver. The one remaining problem was that the GigEVision standard is not publicly available but only to camera manufacturers and some software companies. To the time, fortunately, OpenGigEVision[4] was already in a state where it worked with some cameras and the knowledge of the protocol gathered in OpenGigEVision could be used to get the camera working, as will be seen later.

2.3.3 Optics

The optics of the camera needs to meet the following requirements:

- At about one meter distance, the image should be filled by the target object entirely.

- When the target is far away from the camera, the image must remain sharp without the need for manual focus adjustment.

---

[3] An exception to this statement are several generic device classes, like Human Interface Devices (HID). However, such a universal device class does not exist for high-performance cameras.
[4] The project website is http://gitorious.org/opengigevision, where the source code can be downloaded as well.

The first requirement defines the field of view (FOV) or imaging angles, which were determined to be 40 degrees. The second requirement forces to use a very sensitive camera so the iris size can be reduced to a minimum. The smaller the iris, the sharper the image (for an infinitesimally small iris, one would obtain a pinhole camera again).

For the tests, a Pentax optics has been used with a focal length of $f = 6$ mm. The focal ratio could be tuned to $f/8$, what still provided a sharp image. For some tests, this could even be reduced to $f/16$. Details can be found in chapter 7 starting on page 151.

2.3.4 Mechanical and power

The mechanical properties of the camera were only of minor importance. The camera must be mountable on the EPOS tool plates. The power requirement is likely to be of no concern, because any camera which fits in the requirements defined so far, will not draw too much power. The maximum power draw allowed depends on the operating voltage. The chosen camera is rated 3 W at 12 V.

2.3.5 Final decision

The camera satisfying all the requirements just defined has been chosen to be the Prosilica GC-655. The sensor chip has a large imaging area, what makes it a bit more expensive, but the gain is much more sensitivity to weakly lighted scenes. Most other cameras, which have also been considered, were ruled out mostly because of the interface.

2.3.6 Sensor calibration

A single monocular imaging camera has the following properties, which must be properly addressed in order to achieve the highest possible accuracy:

- **Focal length.** The focal length is a property of the optics describing the location of the point where all rays, which are parallel to the optical axis on one side of the lens, meet. This point must be known (along a few other parameters) for mapping direction vectors to each pixel.

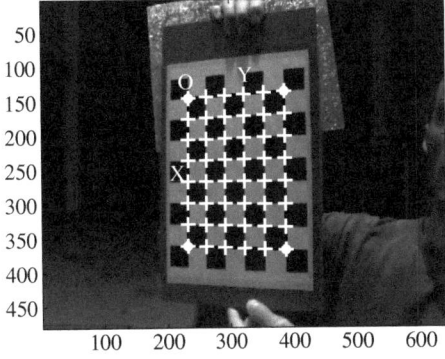

Figure 2.5. Calibration of the CCD camera inside MATLAB. After a few images of the checkerboard have been recorded from different orientations, the four outer corners of the pattern are manually marked. Then, the checkerboard crossing points are automatically traced, and the distortion can be calculated. The X- and Y-axes show the pixel coordinates.

- **Sensor pixel grid.** The individual pixels are of a certain size, depending on the sensor. While more expensive sensors have larger pixels (what makes them much more sensitive to light), inexpensive sensors typically have small sensor areas and, consequently, small pixels. The exact size of a single pixel must be known to retrieve a direction vector for each pixel.

- **Radial distortion.** The problem of mapping a 3D object onto a 2D pixel array can be described with the so-called projection matrix [94]. However, this is only true, as long as the lens is perfect as in theory. In practice, this is not the case. A radial distortion is applied to the projection, which causes imaged points to move on the line between the imaged point and the image center. Depending on the lens, the point can move towards the image center or move away from it.

- **Tangential distortion.** The manufacturing of a camera implies a certain variance of the internal sensor chip positioning. As a consequence, this leads to tangential distortion which must be compensated for maximum accuracy.

- **Skew.** Skew is the deviation from orthogonality of the X and Y axes of the sensor pixel grid. Today's sensors usually have an almost perfect pixel grid, where the skew is zero, so the X and Y axes are orthogonal to each other.

- **Principal point.** The principal point is the point on the sensor, where the optical axis intersects the sensor surface. Due to manufacturing inaccuracies, the principal point can move multiple pixels from the center of the image.

Table 2.2. Intrinsic calibration results for the monocular camera. Values are from the CalTech camera calibration toolbox. 27 images have been used for the estimation.

Property			Estimated*	Expected[†]	Unit
Focal length		X	591.319 ± 4.187	606.061	pixels
		Y	600.138 ± 4.187	606.061	pixels
Principal point		X	339.261 ± 4.052	320.000	pixels
		Y	238.746 ± 5.277	240.000	pixels
Distortion coefficients[‡]		a_0	−0.204438 ± 0.023	0.000	—
		a_1	0.447498 ± 0.121	0.000	—
		a_2	−0.004769 ± 0.002	0.000	—
		a_3	0.002288 ± 0.002	0.000	—
Pixel size		X	—	9.90	μm
		Y	—	9.90	μm

* Uncertainty values are approximately 3σ.
[†] Taken from the camera data sheet.
[‡] The radial distortion is represented by a_0, a_1, while the tangential distortion is represented by a_2, a_3 in accordance with the calibration toolbox.

Since all projection operations rely on this, the principal point is one of the most important properties which must be calibrated.

CCD sensors are well understood and therefore, calibration tools are available. In this subsection, the optical distortion of the camera will be measured. Figure 2.5 shows the application of the calibration tool. It allows calibrating any camera using a checkerboard pattern which has to be imaged from several orientations.

From the knowledge of the pattern (planar surface, known side lengths of all squares) and overdetermining the problem by a large number of squares, one can obtain the optical distortion parameters (details follow).

The calibration was carried out using the MATLAB toolbox from CalTech[4], which is based on Brown's method [20]. For the sake of completeness, there is also a similar tool available from DLR[5]. It is written in IDL and requires the installation of the IDL

---

[4]The CalTech calibration toolbox can be retrieved from http://www.vision.caltech.edu/bouguetj/calib_doc/.

[5]The DLR calibration tool is not a MATLAB toolbox in the sense that it does directly integrate with MATLAB. Instead, it runs independently but provides a convenient MATLAB export. It can be retrieved from http://www.dlr.de/rm/desktopdefault.aspx/tabid-4853/6084_read-9197/

virtual machine[6]. To prevent the necessity of additional effort, the CalTech software was preferred at this point.

The calibration procedure is carried out by first reading the images and manually locating the four outer corners of the checkerboard pattern. Then, the exact locations of all square borders are estimated by the toolbox. From the series of images and the information of how large the squares are in reality, the toolbox calculates the camera parameters by using a numeric optimization approach to minimize the reprojection error (the distances of the calculated corner points to the measured corner points).

When recording the images, it is important to capture a large part of the image space and also to capture the checkerboard pattern at different orientations. This ensures a well-conditioned problem for the solver and hence, a good result.

Table 2.2 shows the calibration results for the camera used in the experiments. Note that the skew angle was not estimated and is assumed to be zero. This is also recommended in the documentation of the toolbox. The axes of today's sensors' pixel grids are sufficiently close to being orthogonal, so this parameter can be neglected.

The distortion coefficients are used as coefficients in correction polynoms (for details, see algorithm 4.1 on page 96 and equation 6.2.5 on page 136). A graphical representation of the optical distortion is shown in figure 2.6. It can be observed that the distortion of the CCD camera is much more complex than the one of the PMD camera (cf. figure 2.11 on page 60) due to the variable focus of the lens.

2.4 PMD sensor

The PMD sensor is an imaging sensor capable of capturing a gray-scale image of a scene along with distance information [81]. To date, several different PMD cameras are available (of which a good overview is given in [39, 105]). Still, the number of choices is not as large as it was for the CCD sensor. Therefore, the sensor selection process is much simpler. The PMDTec Camcube 3.0 was selected for the following reasons:

- **High lateral resolution.** The higher the lateral resolution, the better. The resulting point cloud, which will be used for estimating the pose, highly depends

---
[6]The IDL VM can be retrieved from http://www.ittvis.com/ProductServices/IDL.aspx.

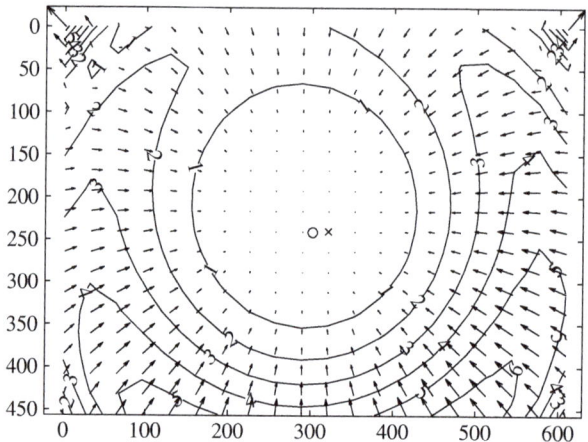

Figure 2.6. Optical distortion of the CCD camera. Numbers indicate imaging deviations in pixels. The cross indicates the geometric image center and the circle indicates the estimated principal point. The X- and Y-axes show the pixel coordinates.

on this parameter. The PMDTec Camcube 3.0 has a resolution of 200 × 200 pixels, which was more than all of its competitors at the time this was written.

- **Suppression of background light.** In space, sun illumination may be present, so all sensors must be able to function even under strong sunlight. An introduction to background light suppression is given in section 2.4.1. Details can be found in [96].

- **Hardware-controlled multiple exposures.** This is very important as it allows using the sensor with larger distances (see section 4.2 on page 105) but also because it allows measurements with multiple different integration times, which can be very useful in high dynamic range scenes, as encountered for example with MLI shielding.

At the time this thesis was written, the only sensor capable of all of the above was the PMDTec Camcube 3.0. An image of the sensor is shown in figure 2.7. In the following, the details of the inner workings of the sensor are given. Table 2.3 summarizes the technical properties of the PMD camera.

At this point, it should be stressed that although the multi-frequency capturing feature can be found in the documentation of the camera, it did not work due to a bug

Figure 2.7. The PMDTec Camcube 3.0 consists of an imaging part and two blocks of LED arrays used to illuminate the scene. *(Image taken from PMDTec/Wikipedia, used in accordance with CC-by-sa 3.0/de license)*

in the firmware, which was fortunately fixed in a later unofficial firmware release provided by PMDtec upon request.

For reproducing the measurements or, for that matter, any attempt of using different frequencies than the one the camera is specified for, a measurement of the modulation signal (see section 7.1.6 on page 157) is highly recommended to ensure valid measurements.

2.4.1 Theory of operation

PMD sensors use modulated infrared light to determine the distance. In practice, the signal used to modulate the LEDs is a square-wave signal [14, page 107]. Due to the low-pass nature of the LEDs, the signal emitted is of a near-sinusoidal shape [109].[4] To simplify further calculations, however, sensor manufacturers assume a sine signal with the same base frequency, neglecting higher frequencies in the signal spectrum. This is what is being correlated with the retrieved signal.

The complete channel design of a PMD pixel is shown in figure 2.8. As light falls onto the pixel, electrons are put into the conduction band of the photosensitive substrate. The electric field induced by the modulation electrodes then creates a charge distribution, which depends on the phase shift between the modulation signal and the light received. This charge distribution causes a low, but measurable voltage at the readout electrodes (which is measured at zero crossings of the modulation signal). Due to noise effects, this voltage must be integrated over time. The time for

---

[4]For an investigation of alternative implementation methods, for example pseudo random noise modulation, see [58].

Table 2.3. Details of the PMD camera (PMDtec CamCube 3). The specifications were taken from the data sheet, except for the integration time limits, which have been determined experimentally.

Property		value	unit
Sensor size	horizontal	200	pixels
	vertical	200	pixels
Lens	focal length	12.8	mm
	field of view (FOV)	40	deg
Modulation frequencies*		18	MHz
		19	MHz
		20	MHz
		21	MHz
Integration time	min	20	µs
	max	50	ms
Repeatability (1σ)[†]		< 3	mm
Operating wavelength		870	nm
Pixel size[‡]		45×45	µm

* The camera comes calibrated for operation at 20 MHz only. Multiple exposure mode can be used with 2 or 4 frequencies. At the same time, integration times must be provided separately for each exposure. Only the mentioned four frequencies are valid and usable.

[†] This value was taken from the datasheet of the camera. For practical accuracy estimation and results, see section 4.1 on page 91.

[‡] Missing in the data sheet; taken from [42].

this integration process can be set by the application. Such a circuit can be seen in [74, page 3], where an OP amplifier-based integrator is shown.

For each pixel, four channels $\theta \in \{0, \pi/2, \pi, 3\pi/2\}$ are implemented [81].[5] The voltage differences between A/D converters A_θ and B_θ are a direct readout of the autocorrelation functions

$$S_\theta = \int_0^t S_{M_\theta}(\tau) \cdot S_L(\tau) \, d\tau \qquad (2.4.1)$$

of the modulation signals S_{M_θ} and the received signal S_L, for some integration time t. The above integrals are measured using two A/D converters each, such that $S_\theta = U_{A,\theta} - U_{B,\theta}$. This is a practical solution for obtaining a fast and symmetric

---
[5] The measurements from the individual channels are performed in a serial fashion but are assumed to have happened at the same point in time due to the relatively small amount of time required per measurement.

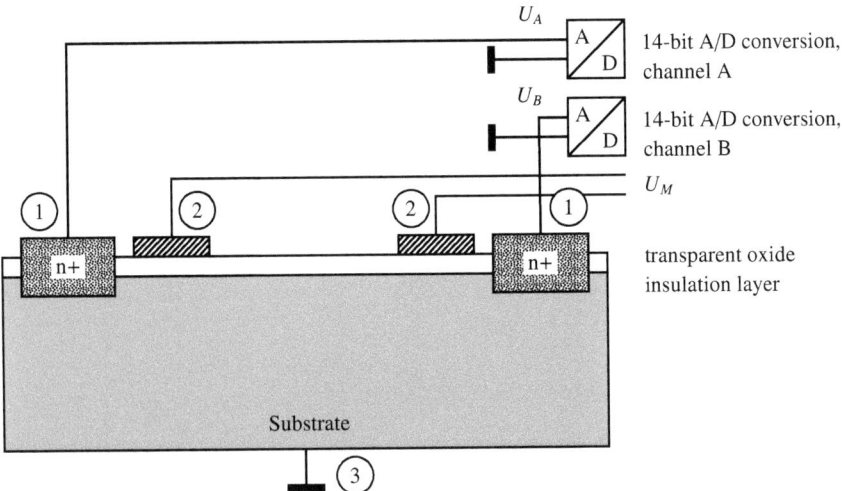

Figure 2.8. The architecture of a single channel of a PMD pixel. The five-terminal structure consists of readout electrodes (1, dotted), modulation electrodes (2) and the substrate ground terminal (3). The oxide layer (white) and the modulation electrodes (hatched) are optically transparent. The A/D converters include a voltage integrator, which is triggered according to the integration time set by the application.

A/D converter in the process (symmetric means, it can read positive and negative voltages). Also and more important, any offset errors present at this position will cancel each other out as well.

The modulation signals are generated for $\theta \in \{0, \pi/2, \pi, 3\pi/2\}$. As a consequence, this allows measuring four different components of the signal: The positive and negative real component, and the positive and negative imaginary component.

Knowing that the modulation signal is sinusoidal, the correlation integral becomes

$$\int_0^t S_{M_\theta}(\tau) \cdot S_L(\tau)\, d\tau = \int_0^t (\sin(\omega\tau + \theta)) \cdot (\psi \sin(\omega\tau + \phi))\, d\tau. \qquad (2.4.2)$$

Here, the modulation signal is represented by a sine term with the frequency embedded in $\omega = 2\pi f_M$. θ is the phase variance of the modulator channels (of which there are four present), ϕ is the phase shift of the measured signal and ψ is the signal strength of the received signal. Solving the above integral yields

$$S_\theta = \psi \frac{t\omega \cos(\phi - \theta) - \cos(\phi + \theta + t\omega)\sin(t\omega)}{2\omega}. \tag{2.4.3}$$

It is now important to see that the four channels will sample a signal at the real and imaginary axes of the complex plane (for each axis, on the positive and negative semiaxes). They are mapped by traversing the complex plane counter-clockwise. For better association, the indices of the autocorrelation functions have already been chosen to represent the corresponding angle. After subtracting the negative real value from the positive one, what yields

$$S_0(t) - S_\pi(t) = \psi t \cos(\phi) - \frac{\psi \cos(\phi + t\omega)\sin(t\omega)}{\omega}, \tag{2.4.4}$$

and the same for the imaginary parts

$$S_{\pi/2}(t) - S_{3\pi/2}(t) = \psi t \sin(\phi) + \frac{\psi \sin(\phi + t\omega)\sin(t\omega)}{\omega}, \tag{2.4.5}$$

it becomes obvious that the dominant part of both terms will be the first sine/cosine terms, while the fraction will practically vanish, because the numerators are bounded by $\pm \psi$ and the denominators only contain ω. Since $\omega \gg 1$ and $t \gg 0$, they can be neglected. What remains are the sine and cosine of the phase angle. As a consequence, the phase angle can be retrieved by calculating

$$\phi = \pi + \arctan2(S_{\pi/2} - S_{3\pi/2}, S_0 - S_\pi). \tag{2.4.6}$$

Note that π was added because of the value range of arctan2, which is shifted from $[-\pi, \pi)$ to $[0, 2\pi)$.

Distance

From here, it follows that the distance can be computed as

$$d = \left(\frac{c}{2 \cdot f_M} \cdot \frac{\phi}{2\pi} - \epsilon\right) \bmod \frac{c}{2 \cdot f_M}, \tag{2.4.7}$$

where f_M is the modulation frequency, c is the speed of light and ϵ is a sensor specific offset to compensate signal run-time delays on the semiconductor. The modulo operator is used to limit the resulting distance to the correct ambiguity interval.

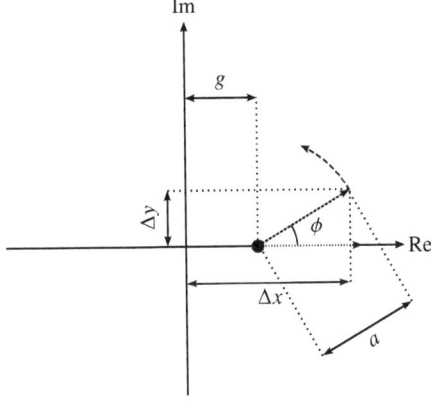

Figure 2.9. Measuring phase, amplitude and DC offset of the returning signal. The dotted arrow (horizontal) is the reference signal and the dashed arrow (sloped) is the received signal. ϕ is the phase shift, g the DC offset and a the amplitude. Equations 2.4.8, 2.4.10 and 2.4.6 follow directly.

Amplitude

The signal strength, i.e., the *amplitude* of the correlated signal can be obtained by determining the length of the signal vector (see figure 2.9) described by its two components $S_0 - S_\pi$ and $S_{\pi/2} - S_{3\pi/2}$:

$$a = \frac{1}{2}\sqrt{(S_0 - S_\pi)^2 + (S_{\pi/2} - S_{3\pi/2})^2}. \tag{2.4.8}$$

Note that at this point, it is being distinguished between the measured amplitude (a) and the *real* amplitude (ψ). The measured amplitude will deviate from the real amplitude and it is impossible to recover the real amplitude. However, it can be approximated (neglecting sensor nonlinearities, especially for very strong signals) by

$$\psi \approx \sqrt{\left(\frac{S_\pi - S_{2\pi}}{t}\right)^2 + \left(\frac{S_{\pi/2} - S_{3\pi/2}}{t}\right)^2}. \tag{2.4.9}$$

All publications regarding the PMD camera and applications always use the measured amplitude (a). The dependency on the integration time is often not explicitly mentioned.

The amplitude is a very good signal strength or signal quality indicator. For example, it is used in the initializer part to find the target by selecting pixels with a sufficiently large amplitude value.

Intensity

For measuring the gray scale image, the total amount of light on the pixel (alternatively, the DC part, or the *intensity*) is

$$g = \frac{U_{A,0} + U_{A,\pi/2} + U_{A,\pi} + U_{A,3\pi/2} + U_{B,0} + U_{B,\pi/2} + U_{B,\pi} + U_{B,3\pi/2}}{8}, \quad (2.4.10)$$

because the variant parts cancel each other out over time, and the only thing remaining is the DC part. This is identical to the inner workings of a plain CCD camera from an architecture point of view.

A high value of g is a bad sign, as it becomes very likely that one of the A/D converters will hit the upper boundary of its value range, causing clipped values. From the readout circuit design, it follows that since the integration time can only be regulated chip-wide and not for each individual pixel, it must be set to the largest possible value which does not cause saturated pixels.

For earlier camera models, saturated pixels could be identified by the property of having a large DC part in combination with a small amplitude [149]. However, recent camera models have a circuitry embedded, which suppresses background illumination [78]. Due to the working principle, which will be explained shortly, one can not rely on the intensity reading any more. Nevertheless, this is not a large drawback, as the camera provides a flag matrix and indicates saturated pixels by itself.

Suppression of background light (SBI)

Newer PMD sensors, such as the one used in this thesis, extend the dynamic range by implementing a circuitry, which prevents the sensor from becoming saturated in the presence of large amounts of unmodulated light (background light). This is achieved by constantly monitoring the voltages at the two potential wells of a PMD channel (elementary cell) as it was shown in figure 2.8.

Immediately after the reset, these are positively charged, as no electrons are accumulated. So the measured voltage is at its maximum. Light puts valence electrons into the conduction band due to the photoelectric effect, and the voltage drops.

Once the voltage of one of the two wells hits a trigger voltage, the SBI circuitry imposes a *correction current* on the triggering potential well and regulates this current

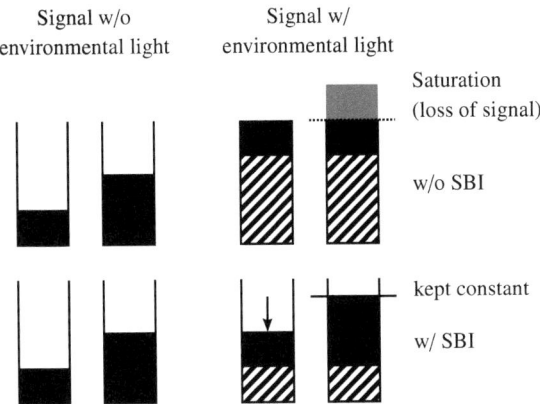

Figure 2.10. Principle of SBI. Charges caused by modulated light are drawn in black, charges resulting from environmental light are drawn hatched.

so that the potential now remains constant. This correction current is simultaneously applied to the other potential well.

When there is no modulated light, both wells will remain static. With modulated light, however, one well is now stabilized, while the other one remains dynamic. Because the well with the larger amount of electrons is kept stable, it follows that the other one will be losing electrons due to the correction current. Consequently, its measured voltage will now increase again, enlarging the signal to noise ratio (figure 2.10).

Unfortunately, besides the benefits, SBI also has a drawback. Because the DC level is not increasing with time any more as it would be the case with SBI off, the DC component can not be used as an indicator for the environmental light any longer. In other words, the intensity reading can not be used for anything meaningful any more. This is also why the intensity channel is not used in this thesis. As soon as strong environmental light is encountered, the reading becomes irrelevant.

SBI operates independently of the camera firmware and can not be controlled in any way by software. However, the Camcube 3.0 camera provides a bit flag for each pixel, which indicates, whether a pixel has activated its SBI circuitry or not. More information about SBI can be found in [78].

2.4.2 Consistency checking

Despite the increasing robustness of the PMD sensor due to SBI, it is advisable to impose a strict consistency checking scheme on the values read from the sensor. This way, erroneous pixels can be marked and either interpolated from neighboring pixels, or be ignored by further processing steps.

At this point, it is clear that the maximum voltage difference the A/D converters can measure, will limit the dynamics of the signal. More precise,

$$g + a < 2^{16} - 1 \qquad (2.4.11)$$

must hold at all times, otherwise the pixel will encounter saturation effects whenever the signal reaches its maximum value. In this case, the pixel should be prevented from contributing to any measurement. Unfortunately, this is only valid for a sensor without SBI, so it is not of practical value for recent sensors.

More usable is the fact that since the signal is periodic and has a sine shape, so the sum of all sampling points must always be zero during one period:

$$\sum_{i=1}^{4} S_{i\pi/2} = 0, \qquad (2.4.12)$$

because this is equal to adding $\Delta x - \Delta y - \Delta x + \Delta y$ (to re-use the terms used in figure 2.9), what should always evaluate to a value near zero (due to noise, it may not be exactly zero, but sufficiently close). In practice, it is sufficient to check that this value is inside a certain confidence interval centered around zero. Deviations from zero indicate pixel saturation and motion artifacts.

Motion artifacts can also be detected by comparing the magnitude of the two measurements of the Δx and Δy values. For a proper measurement,

$$S_{\pi/2} = -S_{3\pi/2} \quad \text{and} \quad S_\pi = -S_{2\pi} \qquad (2.4.13)$$

must hold. If the target object is moving, these identities will not be valid any more. For getting rid of motion artifacts, estimating the distance using only one pair (either $\{S_{\pi/2}, S_{2\pi}\}$ or $\{S_{3\pi/2}, S_\pi\}$) of the four measurement points can be a solution [127]. This, however, is not considered further at this point, as the motion captured by the sensor will be sufficiently slow. However, this may be of great interest when measuring targets spinning at a higher rate.

Table 2.4. Intrinsic calibration results for the PMD camera. Values are from the CalTech camera calibration toolbox. 14 images have been used for the estimation.

Property			Estimated*	Expected[†]	Unit
Focal length	X		289.827 ± 2.541	—	pixels
	Y		293.997 ± 2.567	—	pixels
Principal point	X		98.036 ± 3.192	100.000	pixels
	Y		96.366 ± 3.217	100.000	pixels
Distortion coefficients[‡]		a_0	−0.501351 ± 0.057	0.000	—
		a_1	1.380298 ± 0.688	0.000	—
		a_2	−0.000108 ± 0.002	0.000	—
		a_3	0.004068 ± 0.003	0.000	—

* Uncertainty values are approximately 3σ.
[†] Taken from the camera data sheet. Values not available are denoted by "—".
[‡] The radial distortion is represented by a_0, a_1, while the tangential distortion is represented by a_2, a_3 in accordance with the calibration toolbox.

2.4.3 Sensor calibration

As already introduced and described in section 2.3.6 on page 46, the optics of the PMD camera must be calibrated as well. For the PMD camera, the same calibration procedure was applied, although it is more difficult due to the significantly reduced number of pixels and the resulting loss of resolution. The intrinsic parameters are shown in table 2.4.

Note that the tangential distortion coefficients are slightly smaller than those estimated for the CCD camera. The reason for that is, the PMD camera has a fixed-focus lens, which can not be adjusted, while the CCD camera has a more complex optics, which allows focus and iris adjustments.

The principal point is significantly shifted (figure 2.11). The sensor pixels are usually very small (but still substantially larger than the pixels of the CCD camera due to the additional wiring and per-pixel functionality required). Sensor ICs are mounted with limited precision inside the camera housing, so (small) sensor positioning deviations are common. However, depending on the pixel size, this displacement can manifest in significant error.

As can be seen, the PMD camera does not have visible tangential distortion. This is also suggested by the values in table 2.4 (distortion coefficients a_2 and a_3). Still,

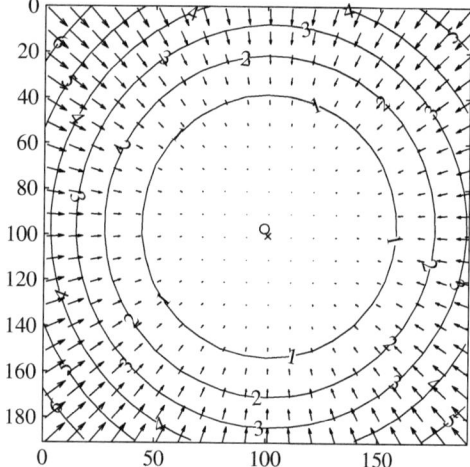

Figure 2.11. Optical distortion of the PMD camera. Both radial and tangential distortion are accounted for in the model. Values in the plot are imaging deviations in pixels. The cross indicates the geometric image center and the circle indicates the estimated principal point. The X- and Y-axes show the pixel coordinates.

the effects of the optical distortion must be accounted for, as the pixel-level offset increases rapidly towards the border of the image. Therefore, the direction vectors (which are used to retrieve a point cloud from a distance image, see section 3.3 on page 88) must also be calculated with respect to these distortion effects.

However, since the PMD camera is mainly used for the distance measurement, the optical calibration is only a part of the complete calibration process. There are still effects related to the PMD chip itself, which will be discussed in detail in chapter 4 (starting on page 91), because it directly relates with the main part of the algorithm.

2.5 Extrinsic sensor calibration

After the intrinsic parameters of the sensors are known, the extrinsic parameters must be determined. The extrinsic parameters describe the exact location and orientation of the sensors, relative to a reference frame. As will be defined later in more detail, the reference frame used here is the satellite body frame, a coordinate frame that is fixed to the satellite structure. The relations between the coordinate frames are shown in figure 2.12.

It is of utmost importance to have an accurate estimate of the sensor orientation, as even small angular displacements can lead to large position estimation errors for objects far away from the sensor. The estimation of the extrinsic parameters

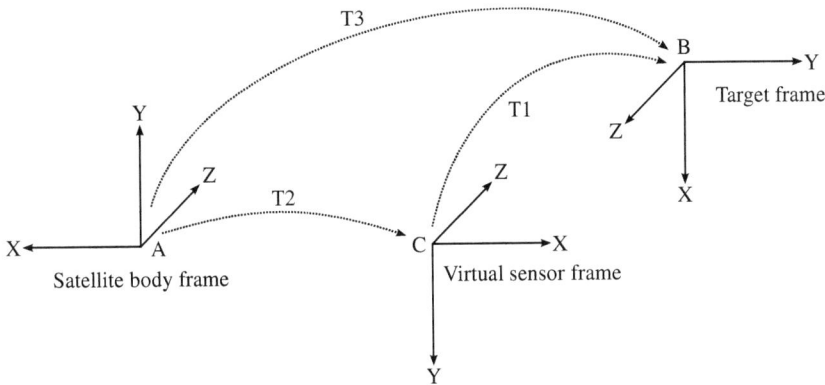

Figure 2.12. Coordinate frames (A, B, C) and transforms (T1, T2, T3) used for the extrinsic calibration. The target frame describes the spatial orientation of the checkerboard pattern and its origin coincides with the upper left corner.

is therefore twofold. First, the camera calibration toolbox is used to retrieve an estimate of the extrinsic parameters, and then the result is refined by a long distance measurement. The quantities most sensitive to orientation error are used to correct the orientation error.

2.5.1 Initial estimation

The first step is the estimation of the extrinsic parameters using the camera calibration toolbox. Two images are used; one for the PMD camera and one for the CCD camera. A checkerboard pattern is mounted to the target robot and moved very close to the sensors, such that two images and reference positions are available (one for each sensor). Then, the camera calibration toolbox is used to extract the extrinsic transformation for both images.

This transformation describes the translation and orientation of the checkerboard pattern relative to a *virtual sensor frame*, which is specific to the calibration toolbox and will later be used to retrieve the transformations to the real sensor coordinate frames. From figure 2.12, the two transformations T1 and T3 are known (T1 from the calibration toolbox and T3 from the EPOS facility) and can be used to estimate T2.

Each transformation will be denoted by a translation vector and a rotation quaternion, where $\mathbf{t}_x$ denotes the translation vector of transformation Tx and $\mathbf{q}_x$ denotes the rotation quaternion belonging to the same transformation.

First, T1 will be determined. This transformation is provided by the camera calibration toolbox, but the translation vector must be converted from millimeters to meters and the rotation quaternion must be constructed from a Rodrigues rotation vector. The Rodrigues form encodes the rotation very similar to a quaternion. The rotation is represented by a three-element vector, where the length corresponds to the angle and the normalized vector corresponds to the vector describing the axis of rotation.

To begin, the direction vector describing the axis of rotation is normalized to the length of 1. Then, the rotation angle can be directly read from the norm of the original vector[4]. The first transformation thus becomes

$$\mathbf{t}_1 = \mathbf{t}_{\text{ext}}/1000 \tag{2.5.1}$$

$$\mathbf{q}_1 = \begin{pmatrix} \frac{\mathbf{r}_{\text{ext}}}{\|\mathbf{r}_{\text{ext}}\|} \cdot \sin(\|\mathbf{r}_{\text{ext}}\|/2) \\ \cos(\|\mathbf{r}_{\text{ext}}\|/2) \end{pmatrix}, \tag{2.5.2}$$

where $\mathbf{t}_{\text{ext}}$ is the translation vector given by the camera calibration toolbox and $\mathbf{r}_{\text{ext}}$ is the Rodrigues rotation vector.

Next, T3 will be defined. The checkerboard pattern (see figure 2.5 on page 47) is mounted with its upper left corner (which will be the origin of the coordinate frame) 12 centimeters to the left and 27.5 centimeters to the top, relative to the mounting plate center. As the mounting plate center coincides with the reference vector provided by the EPOS facility, the translation vector becomes

$$\mathbf{t}_3 = \begin{pmatrix} 0.12 \\ 0.275 \\ 0 \end{pmatrix} + \mathbf{t}_{\text{EPOS}}. \tag{2.5.3}$$

This is a very simple offset, as the mounting plate will not be rotated in any way. The rotation quaternion for T3 can be constructed by inspecting figure 2.12 again. It is created by a rotation of 180 degrees about the Y axis, followed by a rotation of

---

[4] The Caltech camera calibration toolbox handles the Rodrigues vector in a different way than what can be found in literature [40]. Instead of fixing the norm of the vector to be equal to the tangent of half the angle, the norm is set to be directly equal to the angle.

90 degrees about the Z axis. Since coordinate frames are to be rotated, the inverse quaternions of vector rotations must be used (normally, a vector rotation would be performed, however here, the coordinate system is to be rotated). The rotation quaternion thus becomes

$$q_3 = \begin{pmatrix} 0 \\ 0 \\ \sin(\pi/4) \\ \cos(\pi/4) \end{pmatrix} \odot \begin{pmatrix} 0 \\ \sin(\pi/2) \\ 0 \\ \cos(\pi/2) \end{pmatrix}. \quad (2.5.4)$$

Please note that "$\odot$" denotes the quaternion multiplication. Now, in order to retrieve t_2, t_1 is rotated in such a way that it represents the same translation in A as it does in C. Consequently, its coordinate frame must be rotated, first by the inverse of q_1, then by q_3. This yields

$$t_1' = \overline{q_3 \odot \overline{q_1}} \odot \begin{pmatrix} t_1 \\ 0 \end{pmatrix} \odot (q_3 \odot \overline{q_1}), \quad (2.5.5)$$

what describes the same translation as t_1, but now in the satellite body frame A. Then, the missing translation can be determined simply by

$$t_2 = t_3 - t_1'. \quad (2.5.6)$$

The rotation quaternion of T2 is now all that remains to be calculated. It can be retrieved by first rotating the coordinate frame by 180 degrees about the Z axis (to bring the calibration toolbox specific virtual sensor coordinate frame into alignment with the sensor coordinate frames used within this thesis, see figure 2.13), and then applying the rotation quaternion used for obtaining t_1', such that

$$q_2 = (q_3 \odot \overline{q_1}) \odot \overline{\begin{pmatrix} 0 \\ 0 \\ \sin(\pi/2) \\ \cos(\pi/2) \end{pmatrix}}. \quad (2.5.7)$$

The extrinsic parameters of both sensors then are t_2 and q_2. In the following, for better distinction, the parameters of the CCD camera will be denoted by $\tilde{t}_{CCD}$ and $\tilde{q}_{CCD}$. Respectively, for the PMD camera, $\tilde{t}_{PMD}$ and $\tilde{q}_{PMD}$. The tilde is added to

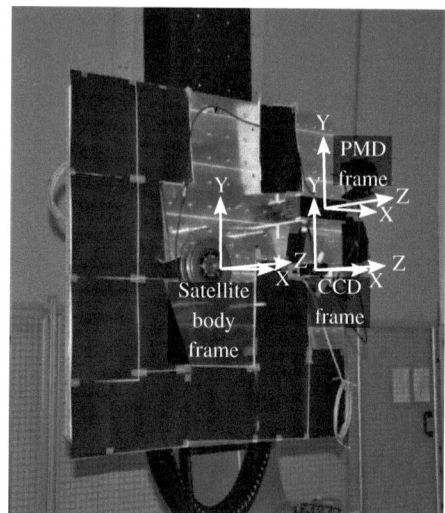

Figure 2.13. Location and orientation of sensor coordinate frames, relative to the satellite body frame (SBF). The Z axis is always pointing towards the target.

indicate that this is only the initial estimate. The coordinate frames used are also shown in figure 2.13 on top of an image of the complete sensor assembly.

Note that the sensor coordinate frames have their origins coinciding with the sensor plane, so their origins have different Z coordinates. Apart from mounting inaccuracies, the coordinate frames are oriented just like the satellite body frame, what is the reference frame used to represent the relative pose of the target object.

2.5.2 Angular refinement

The rotation quaternion just retrieved describes the rotation of the particular sensor relative to the satellite body frame, however, as it is very difficult of estimating the quaternion at small distances, a refinement step has been added which refines the quaternion using a long-range measurement.

In particular, the target position has been estimated using the pose tracker and the extrinsic parameters estimated in section 2.5.1 (page 61), while the target was positioned 22 meters from the chaser. It is of importance to know that at this distance, the rotation quaternion can be corrected with much higher precision when using the translation vectors.

Table 2.5. Extrinsic sensor parameters after the angular refinement step.

Property		PMD camera	CCD camera	Unit
Rotation quaternion components	q_x	0.0094	0.0041	—
	q_y	0.0063	0.0000	—
	q_z	-0.0027	0.0058	—
	q_w	-0.9999	-1.0000	—
Translation components	X	0.2843	0.2783	m
	Y	0.1712	-0.0062	m
	Z	0.1732	0.1379	m

The angular refinement is thus limited to the pitch and yaw angles of the direction vector. This is sufficient, as the roll angle does not have an effect as large as the two caused by the other angles. The pitch and yaw angles of the direction vector are

$$\beta = \arctan\left(t_{m_y} - t_{\text{Ref}_y}\right) / \left(t_{\text{Ref}_z} - 0.225 - t_{\text{CCD}_z}\right) \quad (2.5.8)$$

$$\gamma = \arctan\left(t_{m_x} - t_{\text{Ref}_x}\right) / \left(t_{\text{Ref}_z} - 0.225 - t_{\text{CCD}_z}\right), \quad (2.5.9)$$

where t_{m_x} and and t_{m_y} are the measured X- and Y- components of the translation vector and t_{Ref_x} and t_{Ref_y} are the same translational components, but denote the reference measurement provided by the EPOS facility. 22.5 centimeters are subtracted from the Z component, as this is the thickness of the target mockup. Finally, t_{CCD_z} is the extrinsic translation in Z direction for the CCD sensor determined in section 2.5.1 (page 61). The CCD sensor is used for estimating the angles because of its higher resolution.

Next, two quaternions are constructed from the angles, representing the above angular deviation, $\mathbf{q}_\gamma = (0, \sin(\gamma/2), 0, \cos(\gamma/2))$ and $\mathbf{q}_\beta = (-\sin(\beta/2), 0, 0, \cos(\beta/2))$. Then, given the two previously determined quaternions $\tilde{\mathbf{q}}_{\text{CCD}}$ and $\tilde{\mathbf{q}}_{\text{PMD}}$, the corrected quaternions become

$$\mathbf{q}_{\text{CCD}} = \tilde{\mathbf{q}}_{\text{CCD}} \odot \overline{\mathbf{q}_\beta \odot \mathbf{q}_\gamma} \quad (2.5.10)$$

$$\mathbf{q}_{\text{PMD}} = \tilde{\mathbf{q}}_{\text{PMD}} \odot \overline{\mathbf{q}_\beta \odot \mathbf{q}_\gamma}. \quad (2.5.11)$$

It is assumed here, that the remaining rotational error was common to both sensors. The result of the extrinsic calibration is shown in table 2.5.

2.6 Image processing methods

After the sensors have been introduced, a brief introduction into employed image processing methods is given. As will be seen in chapter 6 (starting on page 129), the CCD sensor will be primarily used to track edges of the target object. For tracking edges, several well-known methods are available. Convolution kernels such as the Sobel filter [134] form the basis of other more elaborate edge detection methods. One of the certainly most well-known edge detection algorithm is the Canny algorithm [22]. Alternative methods utilizing a spectral analysis of the image [73] are intriguing, but have the same problem, that edges are found in the entire image although only very limited regions of interest are relevant. In the envisaged application, known edges should be tracked and the contrast can not be assumed to be stable. Consequently, one must rely on a different method. This is why in the following, the concept of texture segmentation [132] is introduced.

2.6.1 Texture segmentation

Texture segmentation is a computationally expensive, but very powerful image segmentation approach. Instead of relying on very high-contrast edges visible in the image, this works also in areas where the local contrast is much lower, what can be achieved by looking at the local texture properties. The method is therefore capable of segmenting an image or determining edges of objects, even if they can be barely seen. In such a situation, often, humans do not have much difficulty in finding the edge, but all edge detection algorithms do. This is where texture segmentation is very likely to provide a significant improvement.

In the following, the texture segmentation algorithm is briefly summarized, since it is of importance for the 2D part of the algorithm. The algorithm is reproduced from [132], with a few more explanations added where it made sense.

To start, the problem is first reduced to determining the border between two different textures on a scanline (for an explanation, see definition 6.3 on page 137). As a result, the scanline contains pixel intensities S_i. A sequence of pixels provides a sequence of intensities, for example S_1^n. When assuming that the entire scanline has been produced by two different texture-generating processes T_1 and T_2, and their border is at the index c in the scanline, then when determining c (as it is unknown), the probability of the border being at the position c is

$$\text{P}(\text{border at } c|S_1^n, T_1, T_2) = k \cdot \text{P}(S_1^c|T_1) \cdot \text{P}(S_{c+1}^n|T_2), \tag{2.6.1}$$

where k is a normalization constant. The problem is, however, that T_1 and T_2 are also unspecified. Therefore, a generalization must be made: The scanline is now considered to have been generated by *some* texture-generating process. As a result, the probability terms are replaced by integrals over all possible texture-generating processes. For an unknown texture, this leads to

$$\begin{aligned}\text{P}(S_1^c) &= \int \text{P}(S_1^c|T) \cdot \text{P}(T) \, \text{d}T \\ &= \int \text{P}(S_c|T) \cdot \text{P}(S_1^{c-1}|T) \cdot \text{P}(T) \, \text{d}T \\ &= \text{P}(S_1^{c-1}) \cdot \int p_{S_c} \cdot \text{P}(T|S_1^{c-1}) \, \text{d}T.\end{aligned} \tag{2.6.2}$$

The first operation performed was just splitting one factor from a product, while the second operation was an application of the Bayes theorem. Inspecting the integral in the last line yields the insight that this is a sum over the product of the probability of a value and the probability of its occurrence (the probability that the sequence was produced by T), depending on the process T. Therefore, it can be rewritten as

$$\int p_{S_c} \cdot \text{P}(T|S_1^{c-1}) \, \text{d}T = \text{E}(p_{S_c}|S_1^{c-1}), \tag{2.6.3}$$

the expected value of the probability term p_{S_c} when considering all possible texture generating processes. When this is put into the above equation,

$$\text{P}(S_1^c) = \text{P}(S_1^{c-1}) \cdot \text{E}(p_{S_c}|S_1^{c-1}) \tag{2.6.4}$$

suggests a recursive scheme for solving for the probability term on the left side, as long as the expected value on the right can be computed. This is possible, when assuming a uniform prior for T. In practice, this will not always be the case. Still, very often, this assumption approximates reality quite good and even in cases where it does not, the values retrieved are still reasonably stable (this becomes less problematic with increasing length of the scanline, as the brightness distribution will become uniform for an infinite number of pixels).

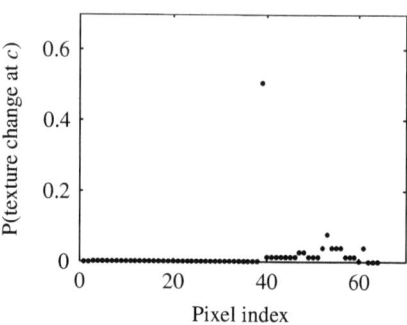

(a) The original image as is it will be processed in the following. The gray line indicates the location and orientation of the scanline.

(b) The probability of texture change determined by the texture segmentation algorithm.

Figure 2.14. The result of the texture segmentation process. First, a scanline is created in such a way that it samples the pixels perpendicular to an edge. Then, the point where a texture change is most likely, is found by evaluating the probability for this texture change. It has a peak at the change point.

For calculating the expected value, one must integrate over all possible combinations of probabilities for the individual intensity values. By weighting the probability value of interest (p_{S_c}) with the normalized probability of T being the texture process describing S_1^{c-1}, one obtains

$$E(p_{S_c}|S_1^{c-1}) = \frac{\int_0^1 \int_0^{1-p_1} \cdots \int_0^{1-\sum_{i=1}^{I-2} p_i} p_{S_c} \prod_{j=1}^{I} p_j^{o_j} \, dp_{I-1} \cdots dp_1}{\int_0^1 \int_0^{1-p_1} \cdots \int_0^{1-\sum_{i=1}^{I-2} p_i} \prod_{j=1}^{I} p_j^{o_j} \, dp_{I-1} \cdots dp_1}, \qquad (2.6.5)$$

where o_j is the number of occurrences of intensity j. Then, solving the above integrals by repeated partial integration yields

$$E(p_{S_c}|S_1^{c-1}) = \frac{o_{S_c} + 1}{c + I - 1}. \qquad (2.6.6)$$

In the original paper, an extension to the above method was presented, which also considers the covariance between a pixel and its predecessor on a scanline. This first-order model follows directly from the just presented 0th-order model. It is not described here in detail, since it does not provide new insights; the calculation is straight-forward with some caution necessary when normalizing the expected value due to the fact that one now has to deal with transition matrices. A typical result of the texture segmentation process is shown in figure 2.14, which was obtained by an implementation of the idea in MATLAB.

The texture segmentation approach just introduced works well when the number of sampled pixels is large compared to the number of possible brightness values. Unfortunately, in the envisaged application scenario, this is not the case. As a consequence, pixel values must be quantized into brightness bins in order to reduce the number of possible brightness values. For details, see section 6.3.2 on page 139.

Typically, a scanline will have 20 to 60 pixels and roughly, the shorter the scanline, the more difficult it becomes to find the edge. Therefore, additional measures for increasing robustness can also be found in the referenced section above. When using high resolution sensors, even better performance of this particular edge detection method can be expected.

2.7 Summary

In this chapter, the EPOS facility along with the PMD sensor and the CCD sensor have been introduced (for further reading about PMD sensors in computer graphics and its applications, see [72]). Also, the optical calibration procedure has been described, which can be applied to both sensors. Here, a state of the art MATLAB toolbox made by the California Institute of Technology was used. With the given compensation for the optical distortion, both sensors can now be used in the following, without having to worry about effects related to the optics.

An in-depth explanation of the distance measurement principle of the PMD sensor was given in order to familiarize with the employed phase shift measurement. Later, this will be of utmost importance for the range extension of the PMD camera. The inner workings of the SBI circuitry have been briefly described as far as an impact on the measurements is to be expected. As the influence of the sun is an important thing to consider for using PMD sensors in space, SBI becomes important as well. Later, measurements will be made which specifically aim for testing the sunlight tolerance of the sensor.

An introduction to the texture segmentation method was given as well, since it is essential for the processing of the CCD image data. The original idea will be extended in section 6.3.2 (starting on page 139) in order to retrieve results even at low image resolution and shorter scanlines, which happen to occur in this application.

Without this modification, the texture information would be represented too sparsely and a reliable edge tracking would become rather difficult. In the measurements performed in chapter 7 (starting on page 151), situations will be encountered where the advantages of this algorithm can be clearly seen (specifically, when almost no light is available).

In the next chapter, the model data preprocessing will be introduced along with the principles and observations behind the architecture of the envisaged sensor data fusion method.

Part II

Algorithm design

3 Off-line preprocessing & Sensor characteristics

Time is a valuable resource in all image processing applications. Especially on satellites, memory and computational power are very limited. Therefore, it is important that only necessary calculations are carried out. The method presented within this thesis is model-based and thus requires model information to be preprocessed in such a way, that the pose estimation runs preferably fast. These preprocessing tasks will be worked out within this chapter.

Also, when inspecting the two chosen sensors, one will soon find very different strengths and weaknesses. In order to ameliorate the weaknesses and make use of the advantages, it is crucial to design a data fusion method specifically for these two sensors. When also incorporating the task or the application, which is pose estimation in this case, even more improvements can be drawn from a proper design of the data fusion process. As a consequence, the outlines of the data fusion method will be developed in this chapter as well (in chapter 6 starting on page 129, this knowledge will be used to design the algorithm in detail).

3.1 Sensor characteristics

In this section, the two sensors will be investigated with regard to estimating the relative pose of objects. This is not to be seen as a quantitative analysis, but rather a way of learning what to expect. Despite the fact that both sensors belong to the optical kind of sensor, there are quite a few differences which will be worked out in the following.

3.1.1 Lateral resolution

As being optical sensors, in both cases, the principal design is the same. The scene is captured using appropriately designed optics, and an imaging chip. While the optics does have an impact on the imaging process, the parameters which are influenced by it are not of major importance at this point (such as field of view, image sharpness, etc).

More important, the properties of the imaging chip must be kept in mind. One major property is the lateral resolution. The number of pixels in the width and height of the sensor determine, how good textures and surface details of objects can be captured.

While the PMD camera has a lateral resolution of 200×200 pixels, the CCD camera has 640×480 pixels. This suggests that surface details and edges visible in the image should be processed using the data of the CCD sensor, as more pixels provide more detail and, consequently, higher accuracy.

3.1.2 Depth perception

When planning to estimate all six degrees of freedom, also the depth (or, in other words, distance of an object from the camera) becomes important. While it is impossible to determine the depth with a single CCD camera without object knowledge (because distance and object size can be changed in such a way that the imaged object is not altered in image space), the PMD camera provides a depth measurement.

Still, when there is information available describing the object of interest, it is also possible to estimate the distance using a single CCD camera only. The problem with that is, however, that the measurement becomes inaccurate with increasing distance [141], because it is an ill-posed problem. At a large distance, even a change in a single pixel can have a large distance measurement error as a consequence.

Therefore, it is advisable to use the PMD camera for measuring the distance. The PMD camera can provide a distance measurement with a fixed accuracy over the whole measurement range. Also, the measurement is stable in the sense that small measurement errors do not propagate to large distance measurement errors, as it would happen in the CCD case.

3.1.3 Frame rate

The frame rate of a sensor becomes important, when motion is to be imaged. Because motion is always represented as a series of single images captured by a sensor, the measurement is always an approximation of reality. As soon as the motion is highly dynamic in terms of velocities, it is of utmost importance that the sensor is fast enough to properly capture it. "Fast" in this context means, the sensor must allow for a high frame rate in order to minimize the time intervals between the captured images.

Both cameras used here can capture images at a frame rate of 10 FPS and even higher. The 10 FPS mark is a rather practical value, which means, in theory, both cameras can do more, but including the interfaces and preprocessing steps, this is very likely to be what can be achieved in real-world scenarios. In space, velocities encountered in rendezvous maneuvers are rather low, so it is unlikely that the frame rate becomes a limit [13, 115].

3.1.4 Noise

Sensor noise poses a problem as it can influence the measurements when not properly addressed. In this case, there are two different noise domains: distance noise in the range measurement of the PMD camera and brightness noise in the image of the CCD camera.

Each noise domain has multiple different noise sources. The noise sources are further of different type in terms of shape and level. For imaging sensors, the most important noise sources are dark current noise (created by thermally produced free electrons) and shot noise (created by the inhomogeneous distribution of photons hitting the sensor).

Noise can be neglected, as long as the signal to noise ratio (SNR) is large enough. For imaging sensors, this is achieved by sufficiently illuminating the scene and using an appropriate exposure time interval. In the remainder of this work, it is assumed, that the SNR is large enough.

The distance noise of the PMD camera has various sources and differs very much from what one would expect. Even more important, the noise amplitude here is too

 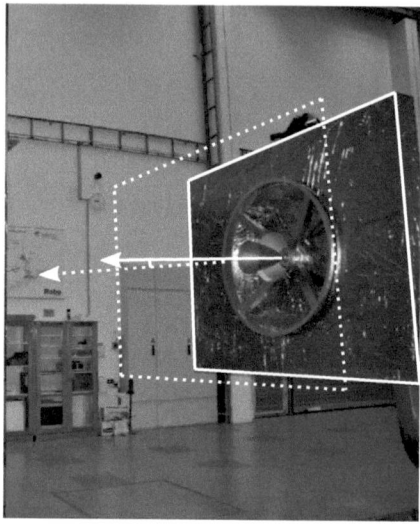

(a) Measurement of the spatial displacement in Z direction. (b) Measurement of the angular displacement for the pitch and yaw angles.

Figure 3.1. Quantities measured by the PMD camera. The distance, as well as the pitch and yaw angles are best measured using the point cloud provided by the PMD camera.

large to be neglected. Thus, dedicated processing is required in order to compensate it. This will be discussed in detail in chapter 4 (starting on page 91).

3.1.5 Separation of the search space

When summarizing the sensor properties, it becomes clear that the PMD camera is very likely to be good at estimating the distance, and also the pitch and yaw angles, what is visualized in figure 3.1. This is because even small changes in these quantities can be measured, when a sufficiently large number of pixels is available.

Comparing this to the CCD camera, the distance can not be measured at all (and must be inferred from the object size, when a model of the target object is available) and the pitch and yaw angles are very difficult to estimate, since small changes translate to even smaller changes in image space.

(a) Measurement of the spatial displacement in X- and Y- direction.

(b) Measurement of the angular displacement for the roll angle.

Figure 3.2. Quantities measured by the CCD camera. Due to the high lateral resolution, the displacement in X- and Y- direction as well as the roll angle are best measured using the CCD camera. The colored overlay parts were drawn manually, as the visualization of inner quantities of S-functions is not possible in MATLAB without long detours.

Consequently, approaches based on monocular vision suffer accuracy losses for these quantities [141]. As a result, the distance and the pitch and yaw angles should be estimated by relying on the depth measurements of the PMD camera.

The roll angle does not lead to changes in the distance but is solely seen in 2D image space. This demands a very high lateral resolution for maximum accuracy. Therefore, as the CCD camera has the larger lateral resolution, the roll angle is best estimated by using data from the CCD camera (figure 3.2b).

After having found displacements for the distance and the pitch, yaw, and roll angles, the position along the image-space axes remains to be solved. It is clear at this point, that this is again a domain for the CCD camera, since motion in these directions will have the largest impact in 2D image space, what again means that the lateral resolution is important (figure 3.2a).

To summarize, the problem is defined as to find the six different displacements measured in the point cloud provided by the PMD camera and suitable image features (e.g., edges) found in the image of the CCD camera. Edges can be tracked in a very robust fashion using texture segmentation, which is why it was decided to implement it that way.

Table 3.1. Summary and reasoning for search space separation.

Pose component	Best representing quantity	Sensor
Pitch angle	Object surface normal vector	PMD camera
Yaw angle	Object surface normal vector	PMD camera
Roll angle	Outer object edges	CCD camera
X position	Outer object edges	CCD camera
Y position	Outer object edges	CCD camera
Z position	Object surface centroid	PMD camera

All of the above statements also hold for rotated objects. They are independent of the object pose, because there are only little displacements which have to be measured. These small displacements can then be optimized to zero by the tracker (see chapter 6 starting on page 129). Table 3.1 summarizes the separation of the search space.

3.2 Model preprocessing

Before the point cloud or edges can be processed, information about the target object must be available in a suitable form. It is assumed that there is a *rough* model of the object of interest available. The definition of what rough means will be given shortly. The model is assumed to be represented as a set of triangles forming surfaces (short: faces). Edges enclose the surfaces.

The model preprocessing task is to extract the surfaces and edges from the triangle set. This is an off-line process, which is performed once in order to obtain the model data required for the on-line processing (section 6.2 and onward, starting on page 134). For an impression of the result, see figure 3.3.

The reason for the preprocessing is the number of possible model sources, which increases, the less constraints are imposed on the model source. With the given algorithm, tracking is possible even with an inaccurate model (with an impact on accuracy, of course).

The model may be retrieved from original CAD files, from a 3D laser scanner object reconstruction and even from manual reconstruction. The point is, however, that the original CAD model of the object to track is *not* required.

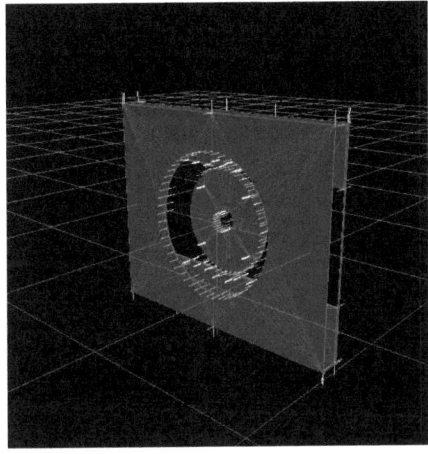

Figure 3.3. OpenGL visualization of the OLEV target model. Irrelevant details, such as the apogee nozzle, have been removed. Some model-specific information is drawn in form of white lines. The gray, 1 × 1 meter grid coincides with the Y axis, which is why it cannot be seen.

In the following, the surfaces of the object are assumed to be planar. Complex objects can still be made up of several planar surfaces of arbitrary shape, so this is not a significant constraint. The problem at this point is more or less the PMD sensor, as the low resolution does not allow for capturing the distance information of highly structured surfaces. The restriction to planar surfaces is thus primarily derived from this specific sensor property.

3.2.1 Surface extraction

The first step in model preprocessing is the extraction of these surfaces[1]. Assuming $\mathcal{T}$ is the set of triangles containing all triangles of the model, where a triangle is represented by three vectors describing its vertices, like $(\mathbf{v}_1, \mathbf{v}_2, \mathbf{v}_3)$, all surfaces can be found by inspecting every triangle $t \in \mathcal{T}$. For each triangle, a surface mapping $f : \mathcal{T} \to \mathcal{S}$ is introduced. A surface $s \in \mathcal{S}$ is defined as a reference point $\mathbf{r}(s)$, a normal vector $\mathbf{n}(s)$ (what defines a plane in the geometric sense), a set of triangles $\mathcal{T}(s)$, a set of edges $\mathcal{E}(s)$ and a set of lines $\mathcal{L}(s)$, which are required for tracking the surface, as will be seen later.

Whenever a surface is instantiated with some of the parameters unspecified, sets are initialized as empty sets. The text will properly define all of the properties, once the points are reached where a particular property is needed.

---

[1] In the rest of the thesis, the word "surface" refers to a finite relation on an infinite 2D subspace of a 3D vector space.

Algorithm 3.1. Surface finding algorithm.

```
Input   : set of triangles T, surface constraint margin ν, normal identity margin ξ
Output  : surface-triangle set mapping T(s), triangle-surface mapping s(t), surface set S
```

1 **foreach** $t \in T$ **do**
2 $\quad \mathbf{n} \leftarrow (\mathbf{v}_2 - \mathbf{v}_1) \times (\mathbf{v}_3 - \mathbf{v}_1)$ // determine the normal vector
3 $\quad \mathbf{n} \leftarrow \mathbf{n} / \|\mathbf{n}\|$
4 $\quad s_0 \leftarrow \epsilon$
5 $\quad$ **foreach** $s \in S$ **do**
6 $\quad\quad \alpha \leftarrow \angle(\mathbf{n}(s), \mathbf{n})$
7 $\quad\quad$ **if** $\|\alpha\| < \xi$ **then**   // Normal vector orientations match?
8 $\quad\quad\quad \mathbf{d} \leftarrow \mathbf{r}(s) - \mathbf{v}_1$
9 $\quad\quad\quad \beta \leftarrow \angle(\mathbf{d}, \mathbf{n}(s))$
10 $\quad\quad\quad$ **if** $\left(\beta > \frac{\pi}{2} - \nu\right) \wedge \left(\beta < \frac{\pi}{2} + \nu\right)$ **then**   // Point is part of the surface?
11 $\quad\quad\quad\quad s_0 \leftarrow s$
12 $\quad\quad\quad\quad$ **break**
13 $\quad$ **if** $s_0 = \epsilon$ **then**
14 $\quad\quad s_0 \leftarrow (\mathbf{v}_1, \mathbf{n}, \emptyset)$ // create a new surface (reference point, normal, triangle set)
15 $\quad\quad S \leftarrow S \cup s_0$
16 $\quad T(s_0) \leftarrow T(s_0) \cup t$
17 $\quad s(t) \leftarrow s_0$

The surface constraint margin ν and the normal identity margin ξ are fuzzifications for allowing noisy measurements of surfaces to qualify as surfaces. In the implementation, setting $\nu = 0.0001$ and $\xi = 0.002$ provided good results.

After the surfaces have been extracted using algorithm 3.1, there may be a large number of surfaces. This may happen in cases where the model contains circular shapes, which need to be approximated by a lot of surfaces and triangles.

To prevent a large number of surfaces from complicating the pose estimation process, the resulting surface set is filtered. Surfaces with a total area (computed from the set of triangles and thus the sum of their individual triangle areas) of less than an area threshold κ are removed. A reasonable value for κ is $0.2\,\mathrm{m}^2$, what is used in the implementation.

3.2.2 Edge finding

Edge finding is the process of marking strips of triangle edges. A triangle t has three edges e_1, e_2, e_3. Each edge may be of a certain type, which will be defined next.

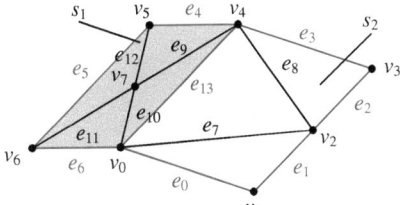

Figure 3.4. Model preprocessing: Inner edges (black), outer edges (gray), vertices (black points) and surfaces (two shades of gray). Valid lines would be $\{e_1, e_2\}$ and all outer edges except e_1 and e_2.

Definition 3.1 (Inner edge) *An edge e of a triangle t, which is itself part of a surface s with a triangle set $\mathcal{T}(s)$ is called an inner edge, when*

$$\forall x \in \mathcal{T}(s) : \forall y \in \mathcal{E}(x) : y \equiv e \Rightarrow \mathcal{S}(y) = \mathcal{S}(e) \wedge \|\mathcal{T}(e)\| > 1,$$

where $\mathcal{S}(x)$ denotes the surface set of the edge x, $\mathcal{E}(t)$ denotes the edge set of the triangle t and $\mathcal{T}(e)$ denotes the set of triangles belonging to the edge e.

This states in a geometric sense that inner edges are the edges, which are only part of a single surface and belong to more than one triangle. Any edge, which belongs to more than one surface or belongs only to one triangle, will be considered an *outer edge*.

Definition 3.2 (Outer edge) *An outer edge is an edge, which is not an inner edge.*

The edges of particular interest here are outer edges. These edges appear either at the border of a surface or at locations where two different surfaces connect to each other. In both cases, these are the edges which can be tracked. More important is, however, that outer edges can be grouped together to a set of connecting edges, which are denoted by the following definition.

Definition 3.3 (Line) *A line is a set of outer edges $\mathcal{E}$, such that*

$$\exists \Psi : \mathbb{N} \to \mathcal{E} : \forall n \in \mathbb{N}, 1 \leq n < \|\mathcal{E}\| : v_e(\Psi(n)) \equiv v_s(\Psi(n+1)),$$

where $v_s(x)$ is the first vertex of the edge x, $v_e(x)$ is the second vertex and Ψ is a suitable edge numbering, which maps a natural number to every edge in the set $\mathcal{E}$.

Note that the above definition explicitly allows straight lines, curved lines and even closed circles, when approximating these shapes with sequences of edges.

Algorithm 3.2. Triangle processing algorithm. For each triangle of all surfaces, all triangle edges are passed to the addEdge subroutine.

Input   : Set of surfaces S
Output : Set of edges $\mathcal{E}$, Outer edge mapping $o : \mathcal{E} \rightarrow \{\text{true}, \text{false}\}$
1 **for** $s \in S$ **do**
2   **for** $t \in \mathcal{T}(s)$ **do**
3   addEdge ($\mathbf{v}(t, 0), \mathbf{v}(t, 1), t$)
4   addEdge ($\mathbf{v}(t, 1), \mathbf{v}(t, 2), t$)
5   addEdge ($\mathbf{v}(t, 2), \mathbf{v}(t, 0), t$)

The extraction of lines and edges starts by finding an outer edge mapping according to definition 3.2. Algorithm 3.2 first determines the set of edges $\mathcal{E}$, along with the outer edge mapping o, which indicates whether an edge $e \in \mathcal{E}$ is an outer edge according to definition 3.2 or not. This is done by considering all detected surfaces and their triangles. The surface set S and the triangle sets $\mathcal{T}(s), s \in S$ have been determined in section 3.2.1 (page 79). The addEdge subroutine is shown in algorithm 3.3.

Algorithm 3.3. Model edge processing algorithm subroutine (addEdge).

Input : Starting vertex $\mathbf{v}_1$, ending vertex $\mathbf{v}_2$, triangle t
1 **Procedure** addEdge ($\mathbf{v}_1, \mathbf{v}_2, t$)
2   **for** $e \in \mathcal{E}$ **do**
3   **if** $(\mathbf{v}(e, 0) = \mathbf{v}_1$ **and** $\mathbf{v}(e, 1) = \mathbf{v}_2)$ **or** $(\mathbf{v}(e, 1) = \mathbf{v}_2$ **and** $\mathbf{v}(e, 0) = \mathbf{v}_1)$ **then**
4              $o(e) \leftarrow$ **false**   // *same edge already seen from another triangle of the surface*
5              $b \leftarrow$ **true**   // *memorize this edge was already seen*
6   **break**
7   **if** $\neg b$ **then**   // *unknown edge, not yet mapped*
8          $\mathcal{E} \leftarrow \mathcal{E} \cup (\mathbf{v}_1, \mathbf{v}_2, t)$ // *add new edge to the set of known edges*

Next, the vertex- edge mapping $f : v \rightarrow e$ must be defined. This is accomplished by iterating over all edges and then storing the mapping for each vertex encountered, but limited to outer edges only. The complete procedure is given in algorithm 3.4.

Now that outer edges are marked and the mappings have been set up, the next step is to obtain lines as sequences of edges in accordance with definition 3.3. This is implemented by iterating over all edges found and traversing the model by the vertices from one edge to the next. By calculating the angle between the edges, it is possible to detect corners. Also, straight lines can be detected when inspecting this

Algorithm 3.4. Vertex/ outer edge mapping. For each outer edge, the edge is stored in one of the two references of the corresponding vertex.

```
1  for x ∈ E, o(x) = true do
2      if (e(v(x,0),0) ≠ x) and (e(v(x,0),1) ≠ x) then        // unseen edge?
3          if e(v(x,0),0) = ε then
4              e(v(x,0),0) ← x                                // memorize this edge in the first edge point
5          else if e(v(x,0),1) = ε then
6              e(v(x,0),1) ← x                                // memorize this edge in the second edge point
```

angle. This is summarized and put into formal terms in algorithm 3.5. Have a look at figure 3.5 for a graphical visualization.

Algorithm 3.5. Line extraction part of the model preprocessing algorithm. Lines are formed by parsing all outer edges.

```
1  for x ∈ E do
2      if o(x) and l(x) = ε then           // outer edge and not assigned to any line?
3          l(x) ← ()                        // instantiate new line and assign
4          parseLine (l, x, v(p,0), false)
5          parseLine (l, x, v(p,1), true)
6          L ← L ∪ l(x)                    // add new line to the set of lines
```

The parseLine subroutine traverses the line in one direction, vertex by vertex. In order to retrieve all edges belonging to the line, it must be traversed in both directions. This is why parseLine is called twice with a different direction flag (last parameter). The complete pseudo code listing is provided in algorithm 3.6.

There are two important parameters, which must be set before the algorithm can be used. These are the corner threshold β and the straightness threshold γ. The corner threshold decides when to stop parsing a line, once a large angular deviation is encountered. When the angular deviation is larger than the corner threshold, the line parsing is stopped. In the current implementation, $\beta = 15\,°$.

The straightness threshold is another angle which determines when a line strip is considered not to be straight any more. Whenever the angular deviation is larger then the straightness threshold, the line will still be parsed, but it will not be considered straight any more. In the current implementation, $\gamma = 0.2\,°$.

In the further processing, only straight lines are used for the estimation of the pose. From the architecture of the algorithm, however, it is possible to include any shape

Algorithm 3.6. Line extraction subroutine (parseLine).

Input   : Line to parse l, starting edge e_{start}, next vertex v_{next}, direction indicator $b_{forward}$
1 **Procedure** parseLine ($l, e_{start}, v_{next}, b_{forward}$)
2      $e_{last} \leftarrow \epsilon$ // last edge
3      $e_{current} \leftarrow e_{start}$ // current edge
4      $e_{next} \leftarrow \epsilon$ // next edge
5      $v_{current} \leftarrow \epsilon$ // current vertex
6      $\mathbf{v}_{dir} \leftarrow \begin{cases} v(e_{start}, 0) - v(e_{start}, 1) & v(e_{start}, 0) = v_{next} \\ v(e_{start}, 1) - v(e_{start}, 0) & \text{otherwise} \end{cases}$ // edge direction vector
7      $st(l) \leftarrow$ **true**   // assume all lines to be straight
8   **if** $v(l, 0) = \epsilon$ **then**
9          $v(l, 0) \leftarrow v(e_{start}, 0)$
10   **if** $v(l, 1) = \epsilon$ **then**
11        $v(l, 1) \leftarrow v(e_{start}, 1)$
12   **while true do**
13   **if** $e_{current} \neq \epsilon$ **then**
14   **if** $l(e_{current}) = \epsilon$ **then**
15               $l(e_{current}) \leftarrow l$ // store edge/line mapping
16               $\mathcal{E}(l) \leftarrow \mathcal{E}(l) \cup e_{current}$ // store line/edge mapping
17   **if** $e(v_{next}, 0) \neq e_{current}$ **and** $e(v_{next}, 0) \neq e_{last}$ **then**
18              $e_{next} \leftarrow e(v_{next}, 0)$ // advance edge
19   **else if** $e(v_{next}, 1) \neq e_{current}$ **and** $e(v_{next}, 1) \neq e_{last}$ **then**
20              $e_{next} \leftarrow e(v_{next}, 1)$ // advance edge
21           $e_{last} \leftarrow e_{current}$
22   **if** e_{next} **and** $o(e_{next})$ **then**   // next edge valid and outer edge?
23              $v_{current} \leftarrow v_{next}$
24              $v_{next} \leftarrow \begin{cases} v(e_{next}, 1) & v(e_{next}, 0) = v_{next} \\ v(e_{next}, 0) & \text{otherwise} \end{cases}$ // pick next vertex
25              $\mathbf{v}_{dir}' \leftarrow v_{next} - v_{current}$ // compute current edge direction
26              $\eta \leftarrow \angle(\mathbf{v}_{dir}, \mathbf{v}_{dir}')$ // compute angular deviation
27   **if** $\eta < \beta$ **then**   // angular deviation below corner threshold?
28                  $e_{current} \leftarrow e_{next}$
29                  $\mathbf{v}_{dir} \leftarrow \mathbf{v}_{dir}'$
30   **if** $st(l)$ **and** $\eta \geq \gamma$ **then**   // angular deviation above straightness threshold?
31                      $st(l) \leftarrow$ **false**   // not straight
32   **else**
33   **break**   // corner hit – stop
34   **else**
35   **break**   // next edge is not an outer edge – stop
36   **if** $b_{forward}$ **then**   // memorize start/ endpoint vertices
37        $v(l, 1) \leftarrow v(e_{current}, 1)$
38   **else**
39        $v(l, 0) \leftarrow v(e_{current}, 0)$

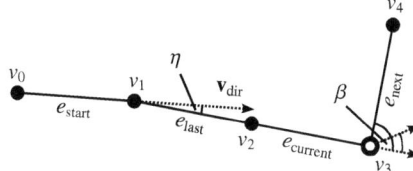

Figure 3.5. Line parsing: from the starting edge, outer edges are traced until a stop condition (such as a corner or no outer edge) is encountered. Edges belonging to the set are stored in the line structure.

of lines without significant additional effort, since the edges are readily available. The reason for this is the reduced computational complexity in the on-line phase of the algorithm.

This completes the line extraction process of the target data. What remains to be done are some preprocessing tasks related to the surfaces of the model.

3.2.3 Surface centroids

The surface centroids are needed for initializing the tracker and aligning the model surfaces with the measured point cloud. Before a centroid can be estimated, it is necessary to obtain a pair of orthogonal base vectors which span the surface. To that purpose, the auxiliary matrix

$$\mathbf{Y}(s) = \Big(\mathbf{b}(t_1) - \mathbf{a}(t_1) \quad \mathbf{c}(t_1) - \mathbf{a}(t_1) \quad \ldots \quad \mathbf{b}(t_n) - \mathbf{a}(t_n) \quad \mathbf{c}(t_n) - \mathbf{a}(t_n)\Big), \quad (3.2.1)$$

where $n = \|\mathcal{T}(s)\|$ is the number of triangles of the surface s, holds the triangle base vectors, which are candidates for the surface base vectors. By using a QR-factorization of $\mathbf{Y}_s = \mathbf{Q}_s \mathbf{R}_s$, $\mathbf{Q}_s$ will hold the orthonormal base vectors spanning the same subspace by definition, which is exactly what is required at this point. The first two columns of $\mathbf{Q}_s$ are thus taken to provide the orthonormal base vectors $\mathbf{b}_x(s)$ and $\mathbf{b}_y(s)$.

Then, for obtaining the centroid of a surface, in general, the first moments of area along two axes must be calculated,

$$\psi_x = \int_A x \, dA \text{ and } \psi_y = \int_A y \, dA. \quad (3.2.2)$$

Since in this case, the surface is not arbitrary (in the sense that only outer polygon corner points are known) but created by triangles, the problem can be reshaped into

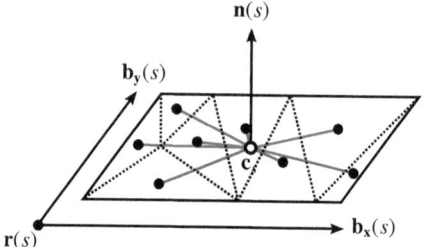

Figure 3.6. Triangle processing. The centroid of the surface (white) is found by calculating the first moments of area of the triangles. By using these moments for weighting the triangle centroids (black), the total centroid is determined.

a summation of the triangle centroids weighted with the distance to the reference axis. After expressing the triangle centroids using the base vectors of the surface s,

$$\tilde{x}_{s,\tau} = ((\mathbf{c}(\tau) - \mathbf{r}(\tau)) \cdot \mathbf{b}_x(s)) \cdot \mathbf{b}_x(s) + \mathbf{r}(\tau) \qquad (3.2.3)$$

$$\tilde{y}_{s,\tau} = ((\mathbf{c}(\tau) - \mathbf{r}(\tau)) \cdot \mathbf{b}_y(s)) \cdot \mathbf{b}_y(s) + \mathbf{r}(\tau), \qquad (3.2.4)$$

the surface centroid can be determined by calculating the first moments of area using the areas of the triangles, what yields

$$\mathbf{c}_s = \frac{1}{A(s)} \begin{pmatrix} \sum_{\tau \in \mathcal{T}(s)} (\mathbf{b}_x(s) \cdot (\mathbf{c}(\tau) - \tilde{\mathbf{x}}_{s,\tau}) \cdot A(\tau) \cdot \mathbf{b}_x(s)) \\ \sum_{\tau \in \mathcal{T}(s)} (\mathbf{b}_y(s) \cdot (\mathbf{c}(\tau) - \tilde{\mathbf{y}}_{s,\tau}) \cdot A(\tau) \cdot \mathbf{b}_y(s)) \end{pmatrix} + \mathbf{r}(p), \qquad (3.2.5)$$

where s is the surface in question, $\mathcal{T}(s)$ is the set of triangles belonging to s, $\mathbf{r}(s)$ is the reference point of the surface (an arbitrarily chosen point which belongs to the surface) and $A(s)$ and $A(\tau)$ are the areas of the surface and the triangles, respectively (figure 3.6).

3.2.4 Principal components of surfaces

At this point, the centroid of each surface $\mathbf{c}_s$ is known along with an orthonormal basis for each surface ($\mathbf{b}_x(s), \mathbf{b}_y(s)$). The problem is that the two base vectors can be arbitrarily rotated about the normal vector of the surface, but for later initialization problems, these vectors must be fixed with respect to the surface.

This can be achieved by computing the principal components (the eigenvectors of the covariance matrix) of the surface in question. The principal components form a base of a subspace of a vector space in such a way that the variance of the data (in this case, the extents of the surface) along the new base vectors is maximized. Therefore, the base vectors just retrieved are replaced by a new set of base vectors obtained from the principal component analysis.

In most cases, the principal components of a point cloud (cf. section 5.2.4 on page 122) or dense set of any kind of data points is required. However, here, the process has to be adapted to a set of triangles.

The first step is to compute the triangle direction vectors, which must be weighted by the triangle area. By projecting the resulting vectors onto the surface base vectors, the matrix

$$\Xi = \begin{pmatrix} (\mathbf{c}_{t_0} - \mathbf{c}_s) \cdot \mathbf{b_x}(s) \cdot A(t_0) & \ldots & (\mathbf{c}_{t_1} - \mathbf{c}_s) \cdot \mathbf{b_x}(s) \cdot A(t_{n-1}) \\ (\mathbf{c}_{t_0} - \mathbf{c}_s) \cdot \mathbf{b_y}(s) \cdot A(t_0) & \ldots & (\mathbf{c}_{t_1} - \mathbf{c}_s) \cdot \mathbf{b_y}(s) \cdot A(t_{n-1}) \end{pmatrix} \quad (3.2.6)$$

is formed. This matrix contains the information of each triangle belonging to the surface, in a way which would allow computing a scaled version of the covariance matrix as $\Xi\Xi^T$. This is the key to obtaining the principal components by using the singular value decomposition. By definition, let $\Xi = \mathbf{U}\Sigma\mathbf{V}^*$. From the definition of the SVD, $\mathbf{U}$ contains the eigenvectors of $\Xi\Xi^T$, while $\mathbf{V}$ contains the eigenvectors of $\Xi^T\Xi$ and Σ contains the singular values of Ξ. Consequently, $\mathbf{U}$ will contain the principal components. The reason for using the SVD is, however, its proven numeric stability and the fact that the decomposition is possible for any matrix (especially non-square matrices) [27].

Continuing with the determination of the principal components, having the two eigenvectors

$$\mathbf{e_x} = \begin{pmatrix} U_{1,1} \\ U_{2,1} \end{pmatrix}, \quad \mathbf{e_y} = \begin{pmatrix} U_{1,2} \\ U_{2,2} \end{pmatrix}, \quad (3.2.7)$$

the 3D principal components are retrieved and memorized as new surface base vectors for the surface s:

$$\mathbf{b_x}(s) = \tilde{\mathbf{b}}_\mathbf{x} \cdot e_{x_1} + \tilde{\mathbf{b}}_\mathbf{y} \cdot e_{x_2} \quad (3.2.8)$$
$$\mathbf{b_y}(s) = \tilde{\mathbf{b}}_\mathbf{x} \cdot e_{y_1} + \tilde{\mathbf{b}}_\mathbf{y} \cdot e_{y_2}, \quad (3.2.9)$$

where $\tilde{\mathbf{b}}_\mathbf{x}$ and $\tilde{\mathbf{b}}_\mathbf{y}$ denote the previously determined base vectors. The new base vectors are now aligned with the principal components of the surface.

For more details about the SVD, have a look at section 5.2.4 on page 122. Here, the same principle is applied in order to determine the principal components of a point cloud. (The results of this section are used to determine the rotation about the normal vector by measuring the angular difference between the principal components of the model and the principal components of a measured point cloud.)

3.2.5 Surface bounding box estimation

Once the new base vectors are available, the bounding box of the surface can be determined. The bounding box is used for the initialization of the tracker. Its dimensions are retrieved by determining the maximum and the minimum of the triangle vertex coordinates in the base frame of the surface, what yields

$$\Gamma(s) = \begin{pmatrix} \max\limits_{\mathbf{x} \in \mathcal{V}(t), \forall t \in \mathcal{T}(s)} (\mathbf{x} \cdot \mathbf{b_x}(s)) - \min\limits_{\mathbf{x} \in \mathcal{V}(t), \forall t \in \mathcal{T}(s)} (\mathbf{x} \cdot \mathbf{b_x}(s)) \\ \max\limits_{\mathbf{x} \in \mathcal{V}(t), \forall t \in \mathcal{T}(s)} (\mathbf{x} \cdot \mathbf{b_y}(s)) - \min\limits_{\mathbf{x} \in \mathcal{V}(t), \forall t \in \mathcal{T}(s)} (\mathbf{x} \cdot \mathbf{b_y}(s)) \end{pmatrix}. \quad (3.2.10)$$

This completes the preprocessing of the model data. What remains to be done in the off-line stage is the calculation of the direction vectors, which map a line of sight vector to each pixel of the PMD camera.

3.3 Direction vector preprocessing

For later use, the pixel direction vectors of the PMD camera are calculated off-line. This allows a fast conversion of a PMD distance image into a point cloud. The direction vector of a pixel with the coordinate vector $\mathbf{p} = (x, y)$ is computed as follows. First, the normalized coordinates (as introduced in section 2.2.2 on page 42)

$$\mathbf{p_u} = \begin{pmatrix} (x - c_x)/f_x \\ (y - c_y)/f_y \end{pmatrix} \quad (3.3.1)$$

are retrieved by subtracting the coordinates of the principal point and dividing by the focal length. The compensation of skew is omitted here, as it is assumed to be zero throughout the document. The remaining part is the inclusion of lens distortion. This is accomplished by applying algorithm 4.1 (page 96) to $\mathbf{p_u}$.

After this step, the direction vector is obtained by complementing the vector with a Z coordinate (for normalized coordinates, the Z coordinate is implicitly defined and assumed to be equal to one). Consequently, the direction vector becomes

$$\mathbf{r_{x,y}} = (\mathbf{p_d}, 1) / \|(\mathbf{p_d}, 1)\|, \quad (3.3.2)$$

where $\mathbf{p}_d$ is the distorted normalized direction vector (the result of applying algorithm 4.1 to $\mathbf{p}_u$). Note that all direction vectors must be normalized to the length of 1 in order to retrieve the correct point cloud during on-line processing.

3.4 Summary

In this chapter, the object model was introduced along with a detailed description of the preprocessing steps, which allow an arbitrary set of triangles to be used as a representation of the object to track. It should be emphasized once more at this point that the triangle set may be retrieved from a detailed CAD model, but this is not a requirement. The algorithm can also handle imprecise models, what is important for situations where a CAD model is not available. As a consequence, an utmost versatile model representation is retrieved, what allows to achieve the most important goal set for the chapter: to have a model representation that allows to track an object with a computational need which is as low as possible. This is achieved by the reduction to surfaces and lines. Existing approaches relying on complete and unprocessed CAD models require on-line processing of a large number of triangles and in combination with least-squares pose residual optimization, this causes the estimation process to take several seconds for a single frame [138].

The separation of the pose search space was discussed as a consequence of the individual sensor properties. It even became possible to map each one of the degrees of freedom of the relative pose estimation problem to one of the sensors in such a way, that the complete pose can be estimated using both sensors in an advantageous fashion. In contrast to most existing methods based purely on stochastic state estimators such as the Kalman filter, this specific way of processing sensor data allows for avoiding working on ill-posed problems leading to unstable and noisy pose estimates which the filter then would have to smooth – a task, which often fails as the noise characteristics start to change rapidly.

Finally, also some preparations of the initialization problem were made (such as determining the principal components of the model surfaces along with their dimensions), where an initial pose of the target object must be found without prior knowledge. This problem will be discussed in full detail in chapter 5 starting on page 115.

In the next chapter, the preprocessing of PMD sensor data will be discussed. This encompasses compensating for sensor-specific non-linearities as well as the processing of multiple frequency raw sensor values for obtaining long-range measurements.

4 PMD Data preprocessing

The PMD sensor is unique in the sense that it unifies several different strengths of various sensors in a single sensor – high frame rate, depth measurement, and low computational requirements. However, there is still a price to pay in form of sensor-specific non-linearities and side effects. These effects are too significant to be neglected. Since the correction of the effects has an impact on the performance of the following processing steps (especially for the range extension), known calibration methods have to be modified and adapted under these circumstances.

Another problem is the limited measurement range of the PMD sensor, what is a consequence of the phase shift measurement principle and the chosen modulation frequency. Since a typical application in space will require a larger measurement range, a method is proposed to extend it, while not relying on the scene geometry as existing approaches do.

4.1 Measurement error compensation

The following is a list of known effects, which must be considered when working with a PMD camera. Please note that since this work has pose estimation as its subject, an extensive calibration of the PMD camera is omitted, except for the major sources of error.

- **Fixed Pattern Noise.** Because every PMD pixel consists of multiple functional parts and a relatively complex processing path, there are several error sources, which contribute to a per-pixel static offset, such as wiring, capacities and inductivities. Fixed pattern noise is the union of all pixel-specific constant offsets.

- **Amplitude dependent distance offset.** Due to the non-linearity of the elementary PMD cells, the amplitude of the signal may have an effect on the phase shift, what translates into a distance measurement error.

- **Wiggling effect.** This is an effect which can be seen when gradually changing the distance of an object while looking at the distance. The error is caused by the LED signal not being exactly sinusoidal due to high-order harmonics remaining in the signal, what is a consequence of the fact that the LED drivers use a rectangular signal. The deviation is not considered during the on-chip correlation.

- **Integration time dependent distance error.** This is an effect which was mentioned by [109] and later confirmed numerous times. The error is characterized by a constant offset which depends on the integration time in a non-linear way. It is probably caused by the non-linear coupling of the read-out voltage and the charge amount of the potential wells of a PMD cell, similar to the amplitude-related offset.

- **Intensity-related distance error.** Environmental light can have an effect on the measurement. The effect can be reduced especially for older sensors, which do not have SBI circuitry. For newer sensors, the compensation becomes difficult if not impossible due to SBI.

- **Motion artifacts.** When the scene captured by the PMD camera contains moving objects, motion artifacts can occur. This effect manifests prominently near object edges and can be observed especially with high velocities. It is caused by the assumption, that the distance remains constant during the sequential measurement of the raw values, which is of course not the case for a moving object.

- **Multipath interference.** Whenever the scene contains a part where two surfaces are inclined in such a way that the angle between the surfaces in the direction of the camera becomes too small, modulated light of the emitters can be reflected from one surface to the other and contribute an error to the measurement.

- **Jump edge effect ("flying pixels").** Pixels imaging edges of objects often tend to have an unreliable distance measurement. The distance measured is often in the middle of the air, instead of being at the object edge. This is caused by more than one contiguous surface contributing to the measurement of a single pixel.

Calibrating PMD cameras has been done to a fair extent and the above effects have been properly addressed in various papers, such as [46, 86, 88, 90, 108, 125, 149], to mention a few. Several of the effects described in the following are related to the partially erroneous integration of the signal, what is caused by a non-linear correspondence of PMD elementary cell electrical potential to photons. Despite the fact that it has already been proposed to use non-linear integration characteristics [2, Chapter 4], the problem remains yet to be solved in ways presented in the above papers. In this thesis, also a state-of-the-art calibration will be made which is based on these articles.

Apart from the static parameters, one must also keep in mind that the LEDs produce a significant amount of heat. The heat must be dissipated into the environment. It also causes the problem that measurements as well as calibration attempts must be made at defined temperatures where the camera is in thermal equilibrium [127]. Therefore, all measurements and calibrations have been made after 20 minutes had passed since the power-up of the PMD camera. The laboratory is air-conditioned and provides temperature and humidity stabilization.

From the given list of error sources, only the following will be compensated: Fixed pattern noise, amplitude dependent distance offset, the Wiggling effect and the integration time dependent distance error. The other effects are not compensated for the following reasons.

First, motion artifacts become a problem at very high velocities, what is not the case here. For Rendezvous and Docking, these motion artifacts are simply negligible, as motion towards the camera or vice versa is known to have only a small effect and the typical velocities encountered in these scenarios are rather low [13, 115]. For a very interesting attempt of compensating the effect, a promising approach was already made by accessing the raw A/D values of the camera and detecting events, where motion artifacts occur. Then, raw images of the previous frame can be used to correct the measurement [127].

Multipath interference has only a very limited effect on the target measurements performed here. Also, if the target had a much more complex surface structure, it would be very difficult compensating the error due to the structure. Most multipath interference correction approaches assume lambertian reflection properties[1] [45], what does not apply to MLI shielding. Furthermore, the computational need for multipath compensation does not scale well; the contributions of every single surface in the vicinity of a pixel must be considered, what becomes impractical for complex surface structures.

The Jump edge effect is not relevant, because invalid measurements are filtered out by the planar RANSAC algorithm (see section 6.2 on page 134). After the filtering, a sufficient number of pixels is still left for being processed in further steps. Nevertheless, when jump edges have to be filtered in a dedicated fashion, [121] provides an overview of available methods.

The intensity-related distance error is not compensated, as with the SBI circuitry and the fact, that it can not be disabled or controlled in any way, no reliable intensity measurements are possible. The camera used here provides a bit flag indicating activity of the SBI circuitry, however, this would only allow to disable any error correction for individual pixels. Since in this context, highly dynamic scenes are to be expected, it does not make much sense implementing a compensation method which would automatically disable itself for a large amount of pixels. Still, when required, in [86], an adequate compensation method can be found.

4.1.1 Fixed pattern noise compensation

The pixels of the PMD chip are relatively complex compared to those of a CCD sensor, as described in section 2.4.1 (page 51). Therefore, depending on the particular chip at hand, some parts of the processing silicon may have been moved to the border of the chip in order to free up space in the active area of the sensor, causing pixel-dependent effects. This is common to all implementations. Also, due to manufacturing tolerances, each PMD pixel is unique due to slightly different sizes of active surface and wiring.

---

[1]Lambertian reflectance is achieved when a body receiving light is scattering the same amount of light in all directions. An observer will always see the same light intensity, independently of the viewing angle.

Fixed pattern noise summarizes all pixel-specific irregularities, which have an impact on the measurement and do not change over time. In practice, these offsets are compensated directly in the distance measurement by adding a matching offset to the measurement.

To calibrate this offset, the camera has been put on a small tool cabinet. Then, the camera was rotated in such a way that it faces perpendicular to a white wall of the laboratory, which can be assumed to be planar. The distance to the wall was 1 m.

An image sequence (100 frames) was captured and the measurement was averaged over all frames to eliminate zero-mean noise, assuming the remaining offsets are fixed.

For better accuracy, first a central offset retrieved from the four inner pixels is calculated,

$$d_o = \frac{1}{4} \sum_{y=100}^{101} \sum_{x=100}^{101} \hat{\mathbf{D}}[y, x], \qquad (4.1.1)$$

which is then used to obtain the relative distances $\hat{\mathbf{D}}' = \hat{\mathbf{D}} - d_o$. For the next step, the pixel direction vectors must be computed as already briefly discussed in section 3.3 (page 88), given the camera calibration parameters:

$$\mathbf{x}' = \begin{pmatrix} (p_1 - c_1)/f_1 \\ (p_2 - c_2)/f_2 \end{pmatrix}, \qquad (4.1.2)$$

where p_1, p_2 are the image-space coordinates of the pixel in question, c_1, c_2 are the coordinates of the principal point and f_1, f_2 are the focal lengths in pixels among both dimensions.

Now, it is important to apply the effects of the optical distortion. This is done by algorithm 4.1, which is part of the Caltech Camera Calibration Toolbox. It is an iterative approach which both accounts for radial and tangential distortion.

The normalized pixel coordinate vectors correspond to a 3D vector with $Z = 1$, so the resulting 2D vectors $\mathbf{x}$ are extended by the third element set to one, such that $\tilde{\mathbf{x}} = (x_1, x_2, 1)$. Then, the direction vector norm $\|\tilde{\mathbf{x}}\|$ can be computed for each pixel.

Let the matrix of vector norms be $\mathbf{V}'$. Then, for each element of $\mathbf{V}'$, the vector norm is adjusted, such that the central four pixels are the zero reference again:

$$\mathbf{V} = \mathbf{V}' - \frac{1}{4} \sum_{y=100}^{101} \sum_{x=100}^{101} \mathbf{V}'[y, x]. \qquad (4.1.3)$$

Algorithm 4.1. Iterative algorithm for estimating normalized point coordinates of a given pixel.

```
Input   : Normalized pixel coordinate vector x'
Input   : Radial distortion coefficients k
Input   : Tangential distortion coefficients t
Output  : Normalized pixel coordinate vector x including the optical distortion model
```
1 $n \leftarrow 0$
2 $\mathbf{x} \leftarrow \mathbf{x}'$
3 **while** $n < 20$ **do**
4 $\quad r \leftarrow x_1^2 + x_2^2$
5 $\quad \gamma \leftarrow 1 + k_1 r + k_2 r^2 + k_3 r^3$
6 $\quad \Delta x_1 \leftarrow 2 t_1 x_1 x_2 + t_2 \left(\gamma + 2 x_1^2 \right)$
7 $\quad \Delta x_2 \leftarrow 2 t_2 x_1 x_2 + t_1 \left(\gamma + 2 x_2^2 \right)$
8 $\quad \mathbf{x} \leftarrow (\mathbf{x}' - \Delta \mathbf{x}) / \gamma$
9 $\quad n \leftarrow n + 1$

Finally, the matrix describing the fixed pattern offset of each pixel is $\mathbf{F}_+ = \hat{\mathbf{D}} - \mathbf{V}$. The resulting matrices, both for a modulation frequency of 18 and 20 MHz are shown in figure 4.1. The distance offset is shown here in meters.

As can be seen in the figure, the fixed-pattern noise seems to have three components. First, the sensor is apparently made up of four discrete blocks, which divide the imaging area into four columns. Second, there is a weak structuring visible, subdividing the blocks into columns of one pixel width. This is probably a result of the internal chip wiring. The last component seems to be a general increase of the measured distance for pixels located near the top of the image, and vice versa for the bottom of the image. The total distance span for the offset depends on the modulation frequency but can be seen to be about 10 centimeters at 20 MHz and about 16 centimeters at 18 MHz.

The irregularity near the left corner at the top is likely to be caused by a pollution of the wall. As this affects only a small number of pixels, its effect will be neglected.

4.1.2 Wiggling effect compensation

One of the most important errors which must be compensated, is the so-called Wiggling-effect. This is a distance-dependent offset on the measurement. It can be characterized as roughly increasing or decreasing with the distance, while at the same time, having an almost sine like periodic part. It is caused by the non-sinusoidal

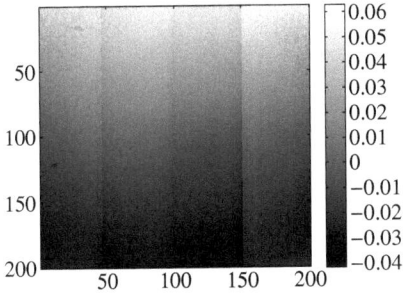 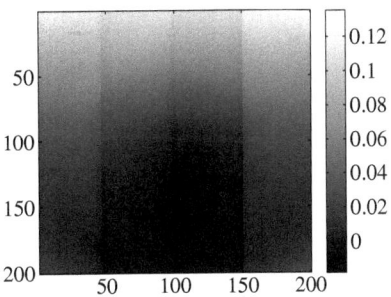

(a) Measured distance deviations for a modulation frequency of 20 MHz.

(b) Measured distance deviations for a modulation frequency of 18 MHz.

Figure 4.1. Visualization of fixed pattern noise of the PMDtec Camcube 3 for the two modulation frequencies used throughout this thesis. The offset is shown in meters. The X- and Y-axes show the pixel coordinates.

form of the optical signal emitted by the LEDs [78], what can be regarded as being a transformed signal, so that in fact the on-chip correlator is performing a cross-correlation, to which the theorems stated in section 2.4.1 (page 51) do technically not apply any more.

There have been several investigations of the problem, which have lead to different approaches for its compensation, for example by fitting B-Splines [90]. Once the B-spline parameters are known, the error can be corrected. Other methods are based on the claim that it should be possible to achieve the same result while using a physical interpretation [102]. The following approach will be based on this work.

The difference is that here, it is possible to estimate the offset for the complete unambiguous interval. Pattinson calibrated the PMD camera for the first unambiguity interval only, what makes it impossible to measure distances down to zero. Also, for use with multiple ambiguity intervals, the error model must be a periodic function. Therefore, the following model does not include a linear term but instead introduces additional low-order harmonics in order to capture the shape of the error, what is the difference to the work of Pattinson [102].

Implementing and continuing with this idea, the problem can be narrowed down to determining seven coefficients of a complex trigonometric polynomial, which represents the most important effect-related harmonics of the signal emitted by the

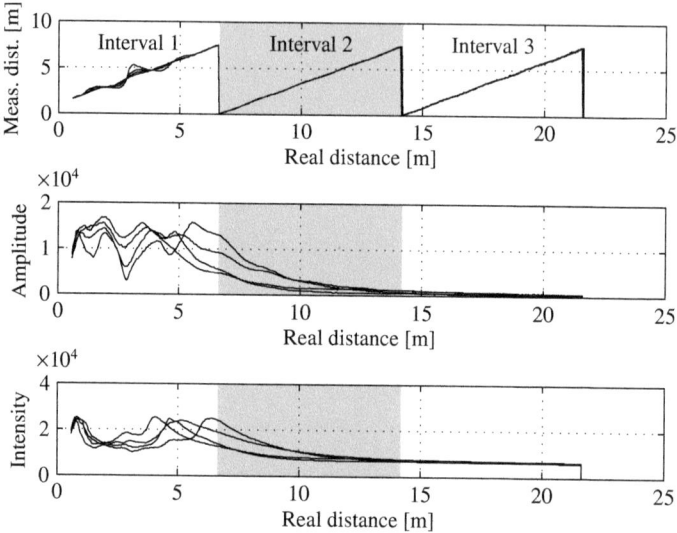

Figure 4.2. Wiggling effect calibration (20 MHz). Four pixels in the center of the image have been measured and are visualized by the four curves, respectively. The gray block shows the ambiguity interval used for fitting the error model.

LEDs[2]:

$$d_+(x) = a_0 + a_1 \cos(2kx) + a_2 \sin(2kx) + a_3 \cos(4kx) +$$
$$a_4 \sin(4kx) + a_5 \cos(8kx) + a_6 \sin(8kx), \quad (4.1.4)$$

where $k = 2\pi/\lambda$ is the *wave number* and x is the measured distance. $d_+(x)$ then provides the offset for correction, once the coefficients a_i are known.

The calibration coefficients a_i are determined by first running a Simulink model on the EPOS facility, what performs a distance sweep from 22 to 1 meter. The PMD camera captures the distance sweep and the data obtained is stored in a file. This file is then processed by a MATLAB script, which processes the four central pixels of the image.

The script performs the following steps: First, it tries to determine the distances where the ambiguity interval overlaps are located. Once known, the modulation frequency and the wavelength are determined. After that, it picks the interval in

---

[2] There is theoretically an infinite number of upper harmonics of such a signal, however, their amplitudes decay rapidly and therefore, only missing harmonics which contribute significantly to the error are considered here.

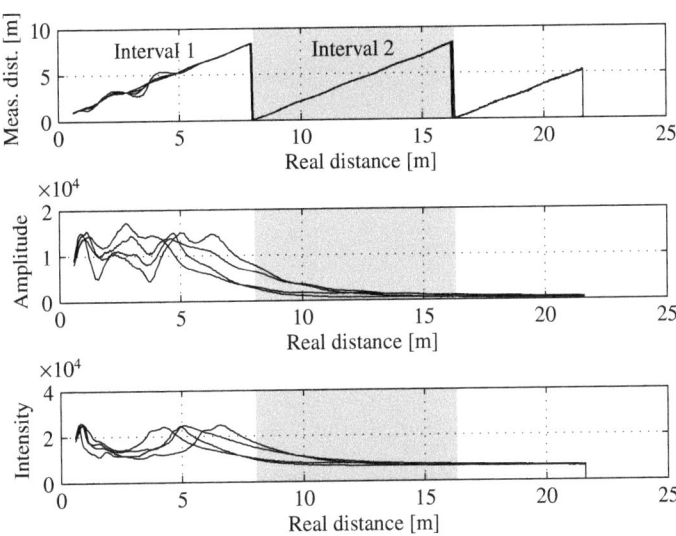

Figure 4.3. Wiggling effect calibration (18 MHz). Four pixels in the center of the image have been measured and are visualized by the four curves, respectively. The gray block shows the ambiguity interval used for fitting the error model.

the middle, determines the static distance offset (the reference distance where a measured distance of zero is read) and then fits the above error model using a least squares approach.

As can be seen in figures 4.2 and 4.3, the ambiguity intervals (see definition 4.1 on page 105) can be automatically identified and a fit can be made once the offset at this position is known. This is particularly useful because calibrating the entire measurement interval at a distance from the camera prevents near-range effects from contributing additional effects to the result, apart from the fact that a distance of zero can not be measured in practice.

The figures also show the measured amplitude and intensity values among the distance. What can be seen from these values is that neither the amplitude nor the intensity seem to have a large effect on the distance measurement, as long as they are kept on a low level. In fact, figure 4.2 suggests that the impact of amplitude changes can be neglected completely as long as it does not lead to saturation effects as can be seen in the first ambiguity interval.

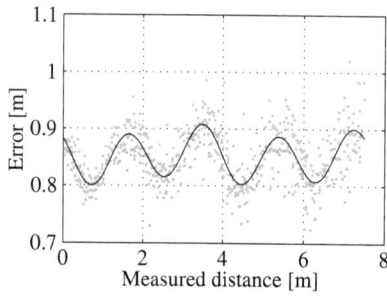
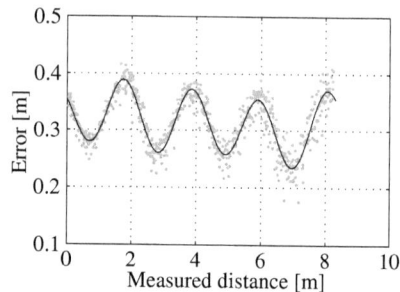

(a) The Wiggling error and its model fit for 20 MHz.

(b) The Wiggling error and its model fit for 18 MHz.

Figure 4.4. Fit of the Wiggling effect error model on data captured by the distance sweep. The gray points show the samples from the measurements. The black line shows the fit of the model.

The result of the calibration is given in figure 4.4. This plot shows the distance error, where measurements are indicated by gray dots and the fit of the error model is shown as a black line. The periodicity is clearly visible. The error model can then be used to compensate for the Wiggling error. Please note that here, the X axis is the *measured* distance and not the reference distance as it was in the plot before. The residuals are given in figure 7.11 on page 166.

As a result, the Wiggling effect is now compensated. However, there are more effects to consider here. The distance sweep for determining the coefficients for the Wiggling effect compensation model has been made at a fixed integration time (2 ms), but measurements may be made with different integration times. Therefore, the integration time offset must be compensated as well.

4.1.3 Integration time offset compensation

It is known from the literature about the PMD sensor, that changes in integration time have indirect side-effects on the distance measurement [39, 86]. In order to compensate this offset, a reference measurement was made, where the integration time was gradually changed from 20 µs to 50 ms. While the integration time was changed, the distance to the reference object was kept constant, so that the influence of the integration time became visible. The offset is approximated for the

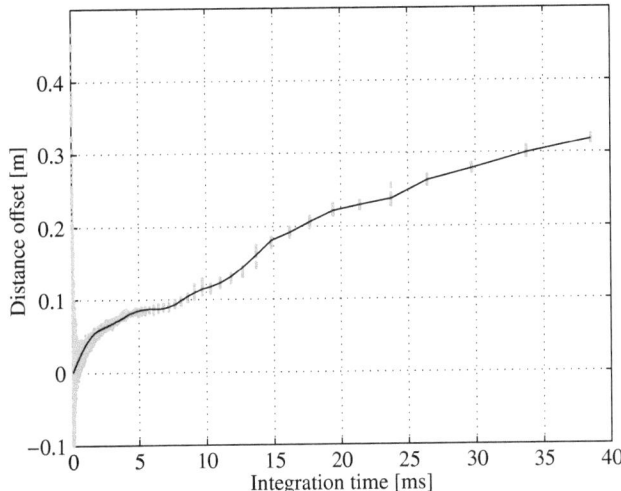

Figure 4.5. Integration time offset for 20 MHz. Gray points show the measurements, the black line shows the modeled approximation using a piecewise second-order polynomial.

compensation using piecewise polynomials of degree 2,

$$t_{+,n}(x) = k_{0,n} + k_{1,n} \cdot x + k_{2,n} \cdot x^2, n \in \mathbb{N}_0, \qquad (4.1.5)$$

as the effect is very complex and can not be reduced to a simpler, more specific model. It was found that 20 interval boundaries distributed uniformly over the value range interval $[20 \times 10^{-6}, 50 \times 10^{-3}]$ are sufficient to achieve a good fit. Figures 4.5 and 4.6 show the results. Note that the offset has been zeroed at 20 µs, because this was the integration time used for the Wiggling effect compensation. As the latter is considered to be the dominating error contributor, all other calibration attempts are made relative to this reference measurement.

At 20 MHz, the offset rises very fast to about 5 centimeters for very short integration times (about 200 µs) and then continues to increase, but having also a continuously changing slope. The entire plot does not show any evidence of saturation. The increasing distance between the clusters of measurements is related to the time it takes the camera to adjust to a new integration time and the sampling grid, which has been designed to have a high density at lower integration times and a low density at higher integration times.

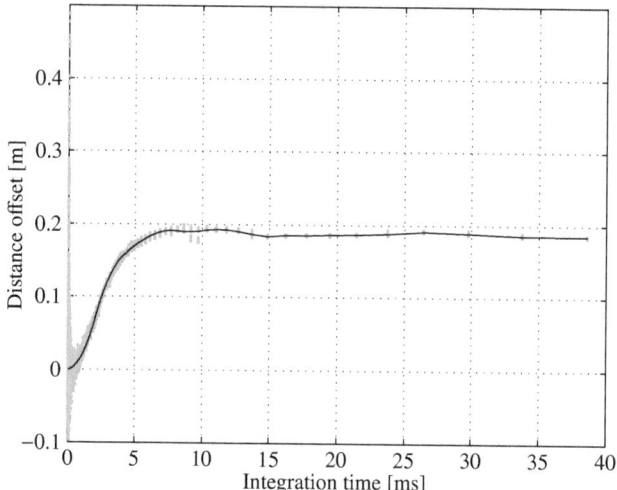

Figure 4.6. Integration time offset for 18 MHz. Gray points show the measurements, the black line shows the modeled approximation using a piecewise second-order polynomial.

At 18 MHz, a saturation can be clearly seen at the offset of about 0.18 m. The plot also shows deviation from the highest offset with increasing integration time. The reason for that is currently unknown.

It is also evident that the effect is very different for the two frequencies. While at 20 MHz, the offset seems to be almost constantly increasing, at 18 MHz it remains at an almost constant level after it has reached it. The cause for this particular difference is also unknown but it is likely to be related to specific optimizations of the PMD chip for a single frequency and in general sensor non-linearities. Integration times higher than 2 ms are uncommon, but may be required in situations where the target has a large distance from the chaser.

In essence, both frequencies must be treated separately when compensating the effect and in case of multi-frequency operation, both models must be available in order to retrieve an accurate overall result.

4.1.4 Amplitude offset compensation

The remaining source of error is the influence of the amplitude on the measurement. As can be seen in figures 4.2 and 4.3 as well, once the distance becomes very small,

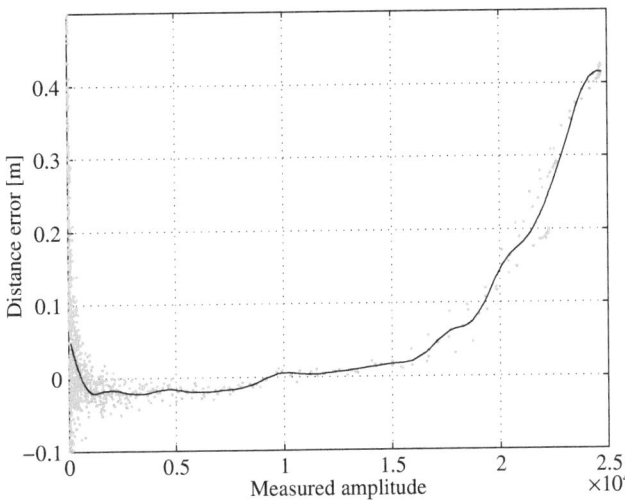

Figure 4.7. Amplitude offset error compensation for the PMD camera at 20 MHz. Gray points denote measurements. The black line is the visualization of the model.

there is very little dampening of the light occurring, because the distance is small. As a consequence, the amplitude of the signal raises. When inspecting the distance plot at this point, the deviation suggests that the amplitude has an effect on the distance measurement once it reaches values near 15,000.

This effect has been investigated in more detail. The target has been positioned at a fixed distance of 5 meters. Then, the four central pixels were read out while the target was being slowly tilted in such a way that the distance of the central pixels remained constant. The amplitude, however, was changing rapidly due to the changing amount of light reflected at the target.

The result for a modulation frequency of 20 MHz is shown in figure 4.7. As was the case with the integration time offset, it is very similar here as well. The effect is non-linear and in this case not monotonically increasing. In fact, it is interesting to see that the offset is positive for very small amplitude values and decreases with increasing amplitude, and at an amplitude of about 950, the offset remains almost stable until at about 8,000, the offset begins to increase.

At an amplitude of around 15,000, also in this plot it can be seen that the error increases rapidly up to 0.4 m. In essence, it is unlikely that this is the symptom of

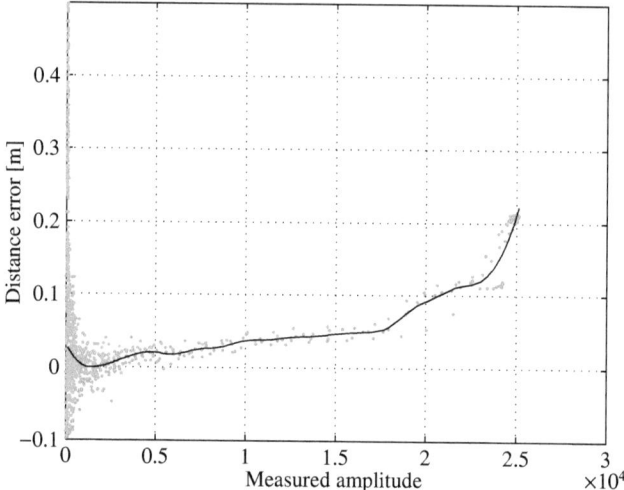

Figure 4.8. Amplitude offset error compensation for the PMD camera at 18 MHz. Gray points denote measurements. The black line is the visualization of the model.

only one source of error – it is probably caused by multiple effects. Unfortunately, details are unknown. At this point it is therefore concluded that the effect must be adequately compensated, especially for high amplitude values due to the magnitude of the offset.

Figure 4.8 shows the result from the same experiment, but for a modulation frequency of 18 MHz. Clearly, there are similarities with the 20 MHz plot. The shift of the offset along the positive Y axis for low amplitude values to zero, the global minimum of the offset and the increasing form for higher amplitude values.

Still, it is interesting to see that the effect is much less significant when measuring at 18 MHz. The offset does not become as large as in the case of 20 MHz (which was about 0.4 m) but instead seems to be limited at about 0.2 m. Nevertheless, it is still too large to be neglected.

To summarize, both figures show that the amplitude-related distance measurement error is probably related to multiple causes. At very low amplitude levels, the noise is very large, as one would expect. However, at the same time the mean value of the samples is also changing. When inspecting both figures, it can be further seen that this is, on a larger scale, frequency-dependent. Therefore, this calibration was made two times for both frequencies.

Another striking part of the figure is the rise of the error at very high amplitude values. It is likely that this is due to sensor saturation. Because of the numerous effects which prohibit using a simple law for approximation, again a set of piecewise polynomials

$$a_{+,n}(x) = k_{0,n} + k_{1,n} \cdot x + k_{2,n} \cdot x^2, n \in \mathbb{N}_0, \qquad (4.1.6)$$

is used for the approximation. For using the fit as a compensation measure, it must be integrated into the existing data preprocessing framework. This has been achieved by adding an additional offset, what forces the measured distances to zero at a *zero offset point*. This point has been selected by reading out the mean amplitude value of the Wiggling error compensation. Then, the distance value of the amplitude offset measurement at this specific point was subtracted. As a result, the offset splines are now zero valued at a reference point which has already been calibrated with the EPOS facility (see section 4.1.2 on page 96 for details).

From now on, all other points with varying amplitude values can be properly handled, assuming the amplitude related offset and the Wiggling effect are uncorrelated errors. Unfortunately, from what is known so far, this is not the case. The given compensation methods reflect the state-of-the-art and provide a fair improvement in accuracy, however, for situations where the amplitude reaches extreme values, the non-linearity of the sensor will become visible again. Nevertheless, as this work concentrates on pose estimation instead of sensor calibration, the PMD sensor specific calibration is regarded to be complete at this point. As will be seen later in chapter 7 (starting on page 151), the remaining distance measurement error does not exceed 11 cm in the worst case (outliers) and remains within 3 cm in the average case.

4.2 Range extension

Before starting with the range extension of the PMD camera, it is necessary to familiarize with the sensor characteristics. The following definition will be of utmost importance for the remainder of this section.

Definition 4.1 (Unambiguous interval) *Given an ideal PMD sensor capable of measuring the phase shift, $[0, \lambda/2)$, where λ is the wavelength corresponding to the modulation frequency, is called the unambiguous interval.*

Sometimes, $\lambda/2$ may be referred to as the *unambiguous range*. This is basically the default measurement range, the PMD sensor is specified for.

The problem that a measurement of the distance is made based on the measurement of a partially ambiguous phase difference is a common topic in image processing [47]. It is encountered in a wide variety of applications [85, 139] and approaches for solving this ambiguity can be typically found under the term *phase unwrapping* in the relevant literature.

When it comes to the PMD sensor, the resulting phase images can also be unwrapped; with a single modulation frequency based on graph cuts [15], constraint/belief propagation [32, 41] and also by using two different modulation frequencies [31] (however, this method relies primarily on phase jumps and uses the second frequency only as an additional clue for the resolution of the phase ambiguity). These approaches are particularly useful for applications in mobile robots due their robustness and the availability of "phase jumps", i.e. distance jumps in the measurement, which are assumed to be caused by a simple wrap-around of the measured phase.

In the scenario discussed within this work, these phase jumps are not available – instead, a single object is seen without any reference to ground or anything else. Therefore, in the following, a different approach is made, which can directly determine the correct distance for any image pixel, independently from the neighborhood. The only limitation is that the range is extended to a *new limit*, so measuring unlimited distances is not possible. When required, however, the method could be easily extended to include object model image-space size constraints, what would allow for ambiguity interval resolution beyond this range. The second limitation is that this is only possible when making an assumption about the noise in the distance measurement, as will be seen later in section 4.2.2 on page 110.

4.2.1 Multi-frequency phase unwrapping

In order to measure distances larger than the unambiguous interval imposed by the modulation frequency of the sensor, in this section, a method is introduced which allows retrieving larger distances. With the given approaches of phase unwrapping, this particular idea has not yet been pursued further due to the lack of an adequate sensor. An existing (but closed) implementation can be found in [14, page 114]. Unfortunately, it is not clear, which method was used in this approach and the publication lacks the details.

Here, the concept will now be developed farther and put to practical use, as the Camcube 3 is likely to be the first sensor available to researchers that allows multi-frequency imaging. Also, specific constraints are derived, which must be followed in practice due to noise effects.

The main idea is, when assuming that there are two different modulation frequencies f_1, f_2 available with corresponding wavelengths λ_1, λ_2 and the sensor can provide images fast enough in such a way that the time between the individual frames can be neglected, then one can obtain the exact distance from the two measurements by solving the following equation system:

$$d = \psi_1 + k_1 \cdot \lambda_1/2$$
$$d = \psi_2 + k_2 \cdot \lambda_2/2,$$

where d is the exact distance, which is to be determined. ψ_1 and ψ_2 are the measured distances from the camera, k_1 and k_2 are distance ambiguity coefficients, i.e., integer factors which define intervals derived from the unambiguous interval as $[k \cdot \lambda_1/2, (k+1) \cdot \lambda_1/2)$ and $[k \cdot \lambda_2/2, (k+1) \cdot \lambda_2/2)$. As a consequence of having two equations and three unknowns[3], the equation system can not be solved in closed form at first glance to provide a solution in practice.

Instead, it is advisable to have a look at what happens with increasing distance (figure 4.9). The measured distance is the result of a measurement process, which could be described by

$$\psi = \left(\tilde{\psi} + \epsilon\right) \bmod (\lambda/2), \qquad (4.2.1)$$

where $\tilde{\psi}$ is the real distance, ϵ is the measurement error and $\lambda/2$ is the unambiguous range again. Having two such measurements with different unambiguous ranges means that the measured values will interchangeably "overflow" with increasing

---

[3]In fact, this will also work for three or more frequencies, however, the time it takes to capture a single frame will increase and it will not solve the problem with the variable count, as there will always be $n+1$ variables in n equations – the one additional variable being the distance.

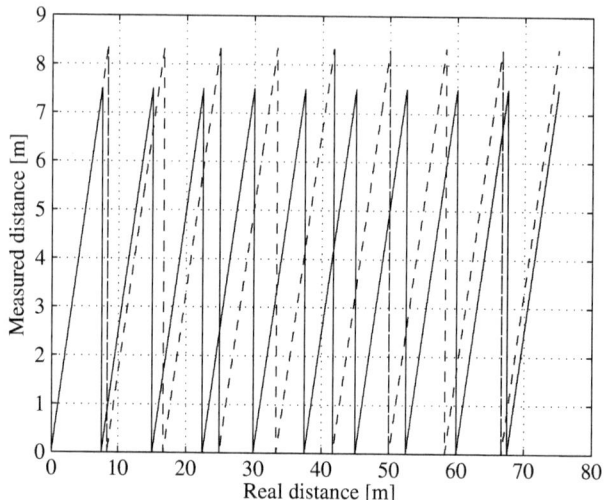

Figure 4.9. Multi-frequency visualization. Solid: measurement with 20 MHz. Dashed: measurement with 18 MHz. Observe the intervals of constant distance between the two signals. On a larger scale, the distances shift depending on the real distance.

distance. This suggests using the difference between both measurements $\Delta\psi$ as an *ambiguity indicator*:

$$\eta = \left\lfloor \frac{\Delta\psi + \lambda_1/2}{\lambda_2/2 - \lambda_1/2} + 0.5 \right\rfloor. \qquad (4.2.2)$$

For most of the time, both measurements will be increasing, so the difference will be constant. But whenever one of the two measurements overlaps, the difference function will instantaneously yield a new value and then again remain constant. The maximum range, which can be measured with this method is thus limited to

$$d_{\max} < \frac{\lambda_1/2 \cdot \lambda_2/2}{\lambda_2/2 - \lambda_1/2}. \qquad (4.2.3)$$

In words, the size of the unambiguous interval of the first modulation frequency times how often the difference between the unambiguous intervals of the two modulation frequencies fits in the unambiguous interval of the second frequency.

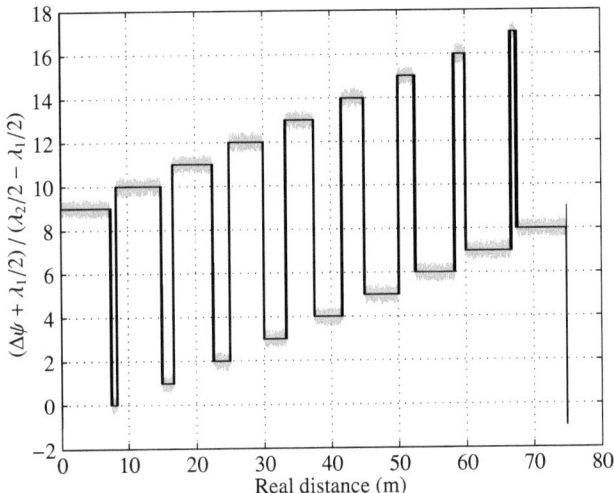

Figure 4.10. Distance difference plot for $f_1 = 20\,\text{MHz}$ and $f_2 = 18\,\text{MHz}$ and artificial noise on the signals. With increasing distance, the deviation between the two signals remains constant at most, but overlaps are clearly visible as sudden changes in value. The black overlay shows values rounded to the nearest natural number.

The nature of the two unambiguous intervals overlapping interchangeably directly implies a schema for obtaining the ambiguity coefficients:

$$(k_1, k_2) = \begin{cases} (0,0) & \eta \in \{0, \hat{\eta}, 2\hat{\eta}\} \\ (\eta+1, \eta) & \eta < \hat{\eta} \\ (\eta - \hat{\eta}, \eta - \hat{\eta}) & \text{otherwise,} \end{cases} \quad (4.2.4)$$

where

$$\hat{\eta} = \left\lfloor \frac{\lambda_1/2}{\lambda_2/2 - \lambda_1/2} + 0.5 \right\rfloor \quad (4.2.5)$$

indicates how often the difference of the two unambiguous intervals fits into the first unambiguous interval. This is essentially equal to half of the number of levels seen in figure 4.10. The main idea behind this formalism is to map the corresponding ambiguity coefficients to each level of the distance difference function.

109

Having determined both coefficients, it is now straightforward to recover the distance from both measurements as

$$d = \psi_1 + k_1 \cdot \lambda_1/2, \text{ or} \tag{4.2.6}$$

$$d = \psi_2 + k_2 \cdot \lambda_2/2, \text{ or} \tag{4.2.7}$$

$$d = \frac{\psi_1 + k_1 \cdot \lambda_1/2 + \psi_2 + k_2 \cdot \lambda_2/2}{2}, \tag{4.2.8}$$

which provide a clean closed-form solution after all. In practice, equation 4.2.6 delivered the best results in the experiments, probably because this was the measurement performed with the frequency the sensor was designed for. Despite the additional calibration attempts performed here, the error was larger for the second frequency. Equation 4.2.8 is specifically not suited for moving targets, but may be advantageous for static scenes.

In the following, this method of estimating the distance from two measurements will be referred to as *direct distance ambiguity estimation* (DDAE).

4.2.2 Noise considerations

Up to now, the theoretical background has been discussed. In practice, one must keep in mind that there is also noise on both signals. Noise causes two effects. First, the constant parts of the difference function will not be constant any more, and become noisy as well. Second, noise will cause *transition zones*, where a measurement might randomly alternate between its upper and lower boundary. This is shown in figure 4.10, and it is also the reason for the rounding in equation 4.2.2.

It is important to understand that knowledge about the signals is very important (noise levels, modulation frequencies and their stability, etc.) and, as figure 4.11 suggests, the distance of all objects must be sufficiently far from zero and the maximum distance, otherwise noise effects will cause unreliable border effects (visible transition zones). These can be detected by observing pixels over a certain time period and memorizing pixels, which have an ambiguity indicator of $\eta \in \{0, 2\hat{\eta}\}$.

As a practical consideration, the noise present in both measurements must be smaller, in its maximum amplitude, than half of the difference between the unambiguous ranges,

$$\|\epsilon\| < \frac{\|\lambda_1/2 - \lambda_2/2\|}{2}, \tag{4.2.9}$$

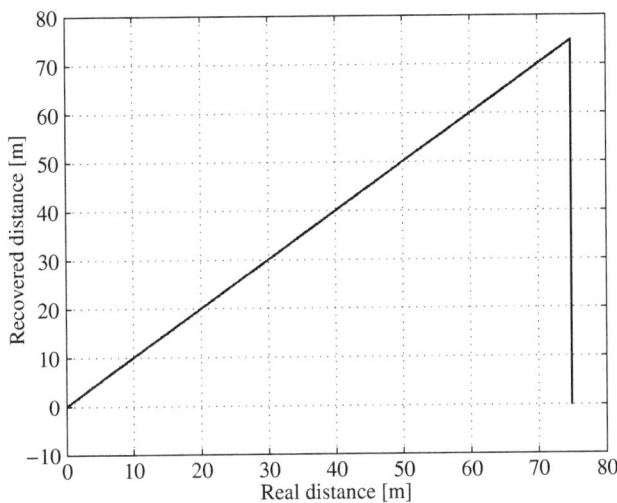

Figure 4.11. Result of the distance recovery: As long as the data does not violate any of the constraints, the distance can be reliably determined. Noise causes border effects – measurements near zero or the maximum possible measurement range should not be trusted. The plot has been made using simulated signals with artificial noise added.

because otherwise, the ambiguity range coefficients k_1, k_2 can not be estimated reliably any more. Note that $\|\epsilon\|$ applies to the sum of the noise components of *both* measurements. So in average, the noise amplitude for a single frequency must not exceed $\|\epsilon\|/2$.

Consequently, DDAF requires properly calibrated sensors and low noise measurements, to achieve best results. In practice, the noise tolerance can be tuned by choosing the modulation frequencies accordingly. When using frequencies which cause equation 4.2.9 to yield a larger value, the noise can have a larger amplitude. However, changing the frequencies and especially making the difference between them larger will decrease the resulting extended measurement range, as equation 4.2.3 indicates.

Thus, one must choose the two frequencies very carefully. In most cases, one will estimate the noise amplitude $\|\epsilon\|$ and then choose the frequencies from the result. Assuming the first frequency is chosen to match best the sensor characteristics, the second frequency can be found by evaluating the noise constraint from

Table 4.1. Chosen modulation frequencies and practical consequences.

Property	Value	Unit
Modulation frequency #1	20	MHz
Modulation frequency #2	18	MHz
Resulting extended measurement range	75	m
Tolerable total noise amplitude ($\|\epsilon\|$)	0.4167	m

equation 4.2.9, which – after some basic calculation – yields

$$f_2 < \frac{c}{c/f_1 + 4 \cdot \|\epsilon\|} \vee f_2 > \frac{c}{c/f_1 - 4 \cdot \|\epsilon\|}. \quad (4.2.10)$$

For the measurements performed for this thesis, the chosen frequencies among more details on the noise tolerance are shown in table 4.1. By using these values, the maximum range measured on the EPOS facility is well below the extended measurement range. Of course, for extending the measurement range further and even for 75 m, other problems must be solved. The LED emitters will very likely not be able to provide enough light intensity for a measurement this far. Following [14, page 114], for distances up to 120 m, an output power of about 8 W is required.

Finally, the entire idea can also be seen as using the camera with a *synthetic modulation frequency* obtained from the difference of the two frequencies used (20 − 18 = 2 MHz). The unambiguous interval is equal to half of the wavelength of this synthetic modulation frequency (150/2 = 75 m). In physics, the related effect is called beat (commonly found in audio contexts).

However, as there are two independent measurements at different times with different noise patterns involved, the former introduction is being considered the correct description of the measurement process. The beat effect was mentioned here due to similarities and deeper understanding.

4.2.3 Noise/ range trade-off

As became clear in the previous subsections, there is a trade-off between the tolerable noise amplitude and the achievable extended distance range. Decreasing the difference between the two frequencies increases the distance range, but this also decreases the tolerable noise amplitude.

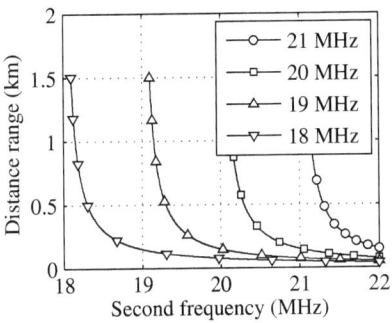 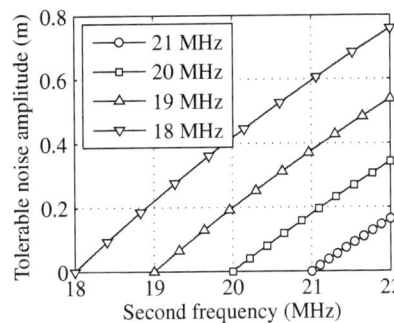

(a) Achievable extended distance ranges for four different primary frequencies.

(b) Noise tolerances for four different primary frequencies.

Figure 4.12. DDAE range/ noise trade-off. Using four different primary frequencies, arbitrary distance ranges are possible. With decreasing noise, significant distance range gains can be achieved, especially for noise levels, which are already very low.

Fortunately, the tolerable noise amplitude scales linearly, while the achievable distance range scales with a quotient. Especially for already very small frequency deviations, even small changes, which decrease the difference further, lead to large gains in distance range and only small changes in the tolerable noise amplitude (figure 4.12).

What can also be seen from the plots is the interesting effect, that the choice of frequencies is irrelevant for the noise margin and the achievable distance range – the only parameter influencing both quantities is the *difference* between the two frequencies. So in effect, the method can be applied to *any* sensor optimized for *any* frequency.

Another conclusion at this point is that a suitable light source (such as a laser) can be used for long-range measurements, as long as the phase shift can be measured accurately enough. The laser provides two advantages: first, it does not have a broad, continuous spectrum, so the infrared filter of the camera can become extremely narrow (eliminating even more environmental light) and second, with a very small beam diversity, its power is sufficient for large distances (see also section 8.4 starting on page 204).

4.3 Summary

In this chapter, a complete list of typical effects of PMD sensors was given. The ones with the largest impact on the distance measurement are compensated with state-of-the-art approaches. Care must be taken when employing the PMD camera for larger ranges, as especially the compensation of the Wiggling effect requires an adequate error model. Finally, after the compensation of the Wiggling effect, fixed pattern noise, integration time offsets and amplitude offsets, an accurate range image is retrieved also for measurements with more than one modulation frequency. Existing calibration methods do not take multi-frequency measurements into account, as this is something which has not been done yet to a large extent. The proposed error compensation is sufficient to achieve a range error of less than 1 % of the object distance for most of the measurement range, as will be seen in chapter 7 starting on page 151. Still, especially the compensation of measurement error of the PMD sensor has room for improvement.

The measurement range of the PMD camera can be extended by using two or more frequencies; the redundancy of the two measurements can be used to extract information about the real distance. However, this method is a trade-off between measurement noise and range. When the modulation frequencies can be chosen freely, the proposed algorithm can be tuned from maximum noise tolerance to maximum range, depending on the requirements. Compared to other approaches for range extension or phase unwrapping, this approach does not require the detection of phase jumps and works independently of the scene geometry. The range extension method just presented has also been published in [144].

This completes the data preprocessing of the sensors. In the next chapter, the initialization problem is discussed, which is roughly stated as to find an initial relative pose estimate of an object without prior information of its whereabouts.

5 Target acquisition

As will be seen later, the pose estimate is retrieved by iteratively refining a given pose estimate on an image-by-image basis. However, the first initial pose estimate must be available before this process can be started. The largest problem is the fact that this must work in absence of any prior information about the target scene. Before calculating a pose estimate, thus the problem of actually *finding* the target object must be solved. In the following, this initial acquisition and estimation process is discussed, and a solution based on the advantages of the PMD camera will be proposed.

5.1 Related work

In this section, some previously published literature is reviewed. Here, the choice of reviewed articles will be limited to approaches, which aim at tracker initialization. Methods such as these are often envisaged with the following guidelines in mind:

- **It does not need to be fast.** For the actual pose estimation process, the initializer is combined with a fast tracker, which can update the pose estimate at a high rate. The initializer therefore does not need to be fast.

- **It does not need to be accurate.** The tracker, which does the updates of the pose estimate for a given sequence of sensor frames is designed to achieve the accuracy required by the application. The initializer only has to fulfill the requirements of the tracker, which are usually much less demanding.

- **Absence of prior knowledge.** The most important property of the initializer is that it must be able to provide the initial pose estimate without any prior knowledge which may aid in the process.

In the following, a few articles presenting methods which have emerged in the last years are reviewed. Most publications rely on single cameras (often RGB cameras) and laser scanners and target a wide range of applications. Consequently, the methods used depend heavily on the latter.

The problem of finding the initial pose is not limited to space-related applications [30]. Depending on the available sensors, approaches based on the Hough transform are viable [25, 68]. Algebraic surface models [137] are particularly elegant and useful for convex objects but generally lack real-time capability. Downsampling an edge image and then using neural networks for initial pose estimation [151] is even real-time capable but needs extensive off-line processing (neural network training) and is sensitive to occlusion.

In a more general context, template matching [21] is often involved, where a known set of images of the target object is correlated with the image of the sensor. The problem is the large search space and the nearest neighbor search, once relevant matches have been found. The latter has been recently ameliorated with pose parameter hashing [133]. An overview of approximation approaches for finding the nearest neighbors is given in [6].

The principal component analysis (PCA) can be used on the point cloud retrieved from a laser scanner to determine two degrees of freedom first, and then the remaining pose parameters can be retrieved by an exhaustive search in 3D space [63]. The drawback is that this takes several seconds even on powerful hardware.

Feature-based approaches are possible for initializing the tracking process as well [143], despite the fact that they increase the size of the model data set and add additional computational need to the off-line preprocessing phase.

To summarize, initialization methods must always consider the entire image and depend heavily on the sensor and the representation of the model. Some methods take several seconds to compute, while others are designed to be real-time capable. Fortunately, the problem of occlusion is not relevant in the Rendezvous and Docking context (except for self-occlusion).

As will be seen in the rest of this chapter, the PMD camera is the source of a lot of advantages, which facilitate the process tremendously.

5.2 Initial pose estimation

While most related solutions commonly utilize a CCD sensor, here, it is of interest to benefit from the advantages of the PMD camera. As the PMD camera is an active sensor, finding the target without any prior information is much easier than using a single CCD sensor. The CCD sensor is specifically unsuited for this task, as the image it provides depends significantly on the illumination conditions, what is not the case for the PMD camera.[1]

Therefore, in this section, a method designed to retrieve a rough pose estimate solely from the PMD camera – without prior knowledge of the target pose – is presented. The result is used to start up the tracker for a refinement of the pose and high-accuracy tracking (see chapter 6 starting on page 129).

5.2.1 Architectural background

The coarse target acquisition is needed in two different situations. First, whenever the pose estimation process is started, a rough estimate is required before it can be refined by the tracker. Second, whenever the tracker encounters a problem and for some reason loses track of the target, the initializer can be used to re-initialize the tracker. This is visualized by figure 5.1.

When the estimator is activated, it is in the *idle* state. It will remain in this state, until an update request forces it to enter the *coarse acquisition* mode, what is discussed in this chapter. Once a rough pose estimate is available, the estimator switches to the *tracking* state. This state remains active until the estimator is deactivated or the target is lost for some reason (tracker divergence, for example). When the target is lost, the pose estimator switches to the *coarse acquisition* state again.

The coarse acquisition state may also lead to a *failure* state, when the target can not be found or the quality of the estimate does not fulfill all requirements. Once this state has been reached, it can never change unless a manual reset is issued. The reason for this behavior is that the initializer should be able to find the target at all times and not finding the target is considered a severe error condition.

---

[1] In fact, of course, also the PMD camera depends on the environmental lighting, as will be seen in chapter 7 (starting on page 151). But first, this is a consequence of a hardware limitation which can be ameliorated and second, the PMD camera can work in complete absence of environmental light in contrast to the CCD camera.

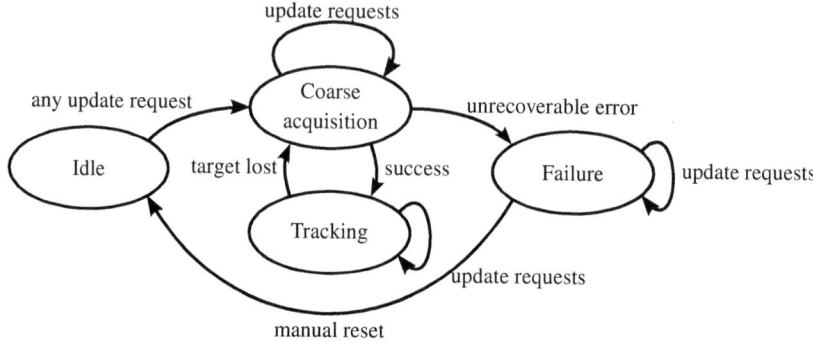

Figure 5.1. Pose estimator state diagram. States are represented by the named ellipses. State transitions are indicated by arrows.

Figure 5.2 shows the structure of the initializer, as it will be constructed in the following. The idea is to retrieve the point cloud of the target, determine the dominant surface (see definition 6.1 on page 134), and then determine the six degrees of freedom for that surface. From there, the pose of the entire object can be calculated.

5.2.2 Image-space coarse acquisition

The acquisition process runs asynchronously. It uses solely the PMD sensor, as it is much easier to find the target here. Most other approaches would use expensive model/template matching algorithms and have to sweep an extremely large search space. Here, an initialization is proposed which is utmost fast and computationally efficient.

The design of the algorithm results from the advantages of the PMD camera: Because the PMD camera is an active sensor, only the target object will produce a significant feedback of the modulation signal, while the background will not. This assumption is always true for situations in space, however, it also works inside the lab, when the background backscatter is sufficiently low.

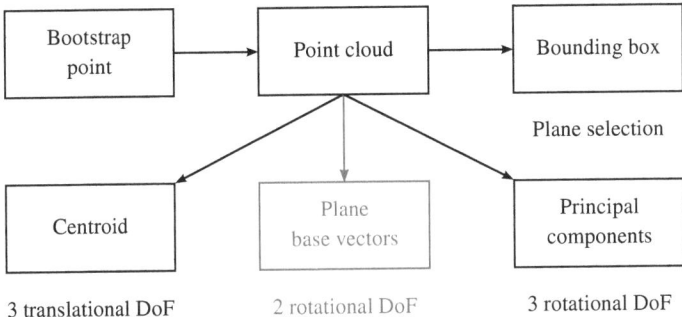

Figure 5.2. Initializer structure. From a bootstrapping point located on the target, a point cloud is constructed. The properties of the point cloud provide the pose estimate.

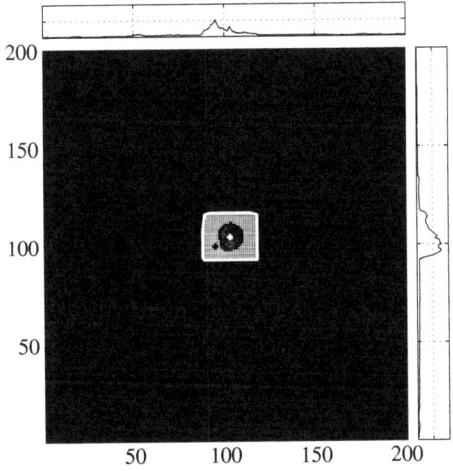

Figure 5.3. Target acquisition visualization. The row and column sums of the amplitude channel are shown on the top and on the right. The global maximum on both dimensions is used to find an inner point of the object (bootstrap point, black cross). Then, object pixels are found to retrieve a point cloud (gray). The centroid (white) is determined from the point cloud. For better readability, the convex hull of the selected pixels is drawn white. The X- and Y-axes show the pixel coordinates.

To begin, the row- and column sums of the amplitude image are calculated (visualized sideways in figure 5.3):

$$\mathbf{u} = \left(\sum_{j=0}^{m-1} \mathbf{A}[0, j], \cdots, \sum_{j=0}^{m-1} \mathbf{A}[n-1, j] \right) \tag{5.2.1}$$

$$\mathbf{v} = \left(\sum_{i=0}^{n-1} \mathbf{A}[i, 0], \cdots, \sum_{i=0}^{n-1} \mathbf{A}[i, m-1] \right), \tag{5.2.2}$$

where m is the number of rows and n is the number of columns of the image. Now the bootstrap point of the target

$$\mathbf{w} = \left(\arg\max_{x \in [0, n-1]} u_x, \arg\max_{y \in [0, m-1]} v_y \right) \tag{5.2.3}$$

is the point obtained by searching the coordinates of the global maximum of the column- and row sums, respectively. The characteristic of this point is that it will always be located on the target and near a spot where the signal is strong. Single pixels with high values will not necessarily influence this measurement, as long as the vicinity does not also exhibit very high amplitude values.

Once the bootstrap point is known, by recursively searching for target pixels along the X- and Y-direction, the surface affinity matrix $\mathbf{X}$ (see definition 6.2 on page 134) is retrieved. This matrix indicates whether a pixel belongs to the target object or not. It is visualized by the gray dots in the middle of figure 5.3. The exact procedure is shown in algorithm 5.1. By calling it with the bootstrap point $\mathbf{w}$ as base position and starting point for the next positions, the recursion is started. After the algorithm has terminated, the surface affinity matrix is available.

The pixels are selected depending on the difference of their amplitude values (α) and their distance values (β). It is assumed that neighboring pixels which are part of a planar surface, have similar amplitude values and a low distance deviation. In the current implementation, $\alpha = 4{,}000$ and $\beta = 0.05$. This works also for rotated objects, due to the fact that only dominant surfaces (see definition 6.1 on page 134) are to be tracked.

5.2.3 Dominant surface estimation

The next step is to calculate the 3D points from the PMD measurements, what is done by multiplying each distance measurement from the distance matrix $\mathbf{D}$ with

Algorithm 5.1. Procedure for determining the surface affinity matrix. From a bootstrap point, all other points belonging to the target are found by recursively searching in all four directions (left, top, right, bottom).

```
1  Procedure scantargetpixels (b, p)
2  if X[p_y, p_x] ≠ 1 then
3     if ‖A[b_y, b_x] − A[p_y, p_x]‖ < α then        // Amplitude difference within threshold?
4        if ‖D[b_y, b_x] − D[p_y, p_x]‖ < β then     // Distance difference within threshold?
5           X[p_y, p_x] ← 1                          // Mark this point as belonging to the target
6           scantargetpixels (p, p + (1, 0))         // recursion: right
7           scantargetpixels (p, p + (−1, 0))        // recursion: left
8           scantargetpixels (p, p + (0, 1))         // recursion: bottom
9           scantargetpixels (p, p + (0, −1))        // recursion: top
```

its corresponding direction vector obtained in section 3.3 (page 88), such that the measured point cloud becomes

$$\mathcal{M} = \bigcup_{\substack{(x,y) \in \Omega_{PMD} \\ X[y,x]=1}} \mathbf{r}_{x,y} \cdot \mathbf{D}[y, x]. \tag{5.2.4}$$

Now given the above point cloud, the task is to determine the target pose. By reducing the problem to the dominant surface, the pose of the target object can be estimated after the surface centroid was retrieved by

$$\tilde{\mathbf{c}} = \frac{1}{\|\mathcal{M}\|} \sum_{\mathbf{x} \in \mathcal{M}} \mathbf{x}. \tag{5.2.5}$$

Then, two (non-orthogonal) base vectors can be determined by evaluating

$$\mathbf{f} = \frac{\sum_{\mathbf{x} \in \mathcal{M}} (\mathbf{x} - \tilde{\mathbf{c}}) \cdot (1, 0, 0)}{\left\| \sum_{\mathbf{x} \in \mathcal{M}} (\mathbf{x} - \tilde{\mathbf{c}}) \cdot (1, 0, 0) \right\|} \tag{5.2.6}$$

$$\mathbf{g} = \frac{\sum_{\mathbf{x} \in \mathcal{M}} (\mathbf{x} - \tilde{\mathbf{c}}) \cdot (0, 1, 0)}{\left\| \sum_{\mathbf{x} \in \mathcal{M}} (\mathbf{x} - \tilde{\mathbf{c}}) \cdot (0, 1, 0) \right\|}. \tag{5.2.7}$$

Given these two vectors, the normal vector of the surface can be retrieved from the normalized cross product $\tilde{\mathbf{n}}(s) = \mathbf{f} \times \mathbf{g} / \|\mathbf{f} \times \mathbf{g}\|$. However, there are several drawbacks when using this method. First of all, depending on the rotation of the surface, singularities can occur (for example, an X-Z surface will produce a division by zero in equation 5.2.7). Since dominant surfaces will have an inclination angle of

45 degrees or less relative to the optical axis, singularities can be excluded but still, there is room for improvement. Therefore, the base vectors of the surface are not estimated right now, but will be estimated in the next subsection, along with the orientation about the normal vector.

So far, five degrees of freedom would have been determined. The surface centroid basically contains the three translational components of the target object, and the surface orientation is defined by the normal vector, which essentially encodes two rotational degrees of freedom.

The only piece of information, which is still missing, is the rotation about the normal vector. In order to retrieve this information, the properties of the surface must be considered. The surface has finite dimensions and more important, a *known shape*. Known shape here refers to the set of triangles known from the model. In the following, it is shown how to retrieve the missing rotational degree of freedom along with more accurate base vectors without the problem of singularities.

5.2.4 Surface orientation

For estimating the rotation about the normal vector, the *principal components* of both the model consisting of sets of triangles and the measured point cloud are determined. Then, by finding the transform that brings both principal components into alignment, the rotation about the normal vector of the surface can be found. Note that this will not work for symmetric targets (worst case: sphere). In this case, *hints* (additional information) must be provided to the initializer in order to retrieve the correct pose estimate. This will be discussed later in section 5.2.5 (page 127).

The first step is to determine the principal components of the point cloud by performing a principal component analysis (PCA). The PCA provides orthogonal base vectors for a real subspace in a vector space oriented in such a way, that the variance of the data is maximized along the principal component vectors. In this case, the vector space is solely spanned by the 3D point cloud $\mathcal{M}$ retrieved from the PMD sensor and the 2D subspace of interest is the visible surface of the target.

Problem reformulation

The principal components of the point cloud are equal to the eigenvectors of the corresponding covariance matrix [10]

$$\mathbf{C} = \frac{1}{\|\mathcal{M}\|} \begin{pmatrix} c_{XX} & c_{XY} & c_{XZ} \\ c_{YX} & c_{YY} & c_{YZ} \\ c_{ZX} & c_{ZY} & c_{ZZ} \end{pmatrix}, \quad (5.2.8)$$

where $c_{XX} = \sum_{\mathbf{p} \in \mathcal{M}} (p_1 - c_1)^2$, $c_{XY} = \sum_{\mathbf{p} \in \mathcal{M}} (p_1 - c_1) \cdot (p_2 - c_2)$ and so on. In practice, this particular problem is best solved by using a slightly different way in order to gain numerical stability – the Singular Value Decomposition. The singular values are a characteristic of a matrix very similar to eigenvalues, but generalized to a larger set of matrices (most importantly, non-quadratic matrices) [27]. The decomposition has the shape

$$\mathbf{M} = \mathbf{U} \Sigma \mathbf{V}^*, \quad (5.2.9)$$

where $\mathbf{U}$ is a unitary matrix containing the so-called left-singular vectors of $\mathbf{M}$ and $\mathbf{V}^*$ is an adjoined unitary matrix containing the right-singular vectors of $\mathbf{M}$. In the following, the adjoint operator will be replaced by the transpose, because all matrices relevant here are real-valued only and in this case, the transpose of a matrix is equal to its adjoint matrix, since the conjugation can be omitted [79, page 334]. Then by definition of the SVD, the left-singular vectors are eigenvectors of $\mathbf{M}\mathbf{M}^T$, and the right-singular vectors are eigenvectors of $\mathbf{M}^T\mathbf{M}$. As a consequence, when rewriting the covariance matrix in the form

$$\mathbf{C} = \frac{1}{\|\mathcal{M}\|} \mathbf{A}\mathbf{A}^T, \quad (5.2.10)$$

where

$$\mathbf{A} = \begin{pmatrix} (\mathbf{p}_1 - \mathbf{c})^T \\ \vdots \\ (\mathbf{p}_{\|\mathcal{M}\|} - \mathbf{c})^T \end{pmatrix} \quad (5.2.11)$$

is a $\|\mathcal{M}\| \times 3$ matrix containing the point coordinates relative to the centroid, the principal components (or, equally, valid eigenvectors of the covariance matrix) can

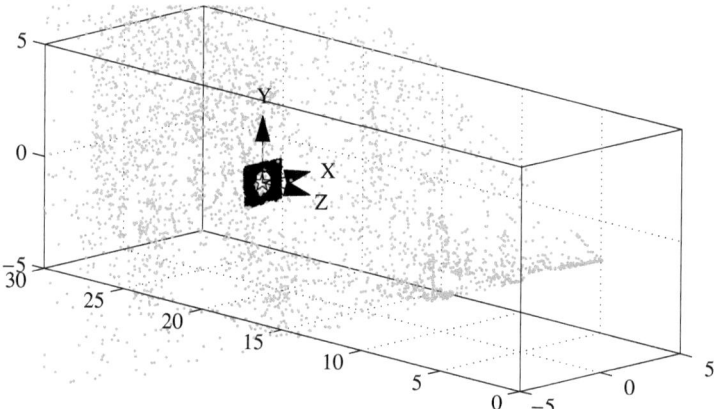

Figure 5.4. Determination of principal components of a point cloud. The target point cloud is visualized by black dots. The two principal components are shown in form of two arrows (X and Y); the other arrow shows the resulting cross product vector. The centroid of the target point cloud is marked as a star. The remaining points of the PMD sensor point cloud (gray) have been sparsified for better readability (only 10 % of all points are shown). All numbers are distances in meters. The camera is located on the right of the image, at position (0,0,0).

be determined by replacing **A** in equation 5.2.10 by the definition of the singular value decomposition of **A**, what yields

$$\begin{aligned} \mathbf{C} &= \frac{1}{\|\mathcal{M}\|} \mathbf{U}\Sigma\mathbf{V}^T \left(\mathbf{U}\Sigma\mathbf{V}^T\right)^T \\ &= \frac{1}{\|\mathcal{M}\|} \mathbf{U}\Sigma\mathbf{V}^T\mathbf{V}\Sigma^T\mathbf{U}^T \\ &= \frac{1}{\|\mathcal{M}\|} \mathbf{U}\Sigma\Sigma^T\mathbf{U}^T. \end{aligned} \qquad (5.2.12)$$

Now the column vectors of **U** are eigenvectors of the covariance matrix **C**. Next, for obtaining the principal components, the first two column vectors of **U** (named $\mathbf{e}_1$ and $\mathbf{e}_2$ in the following) are normalized, such that

$$\tilde{\mathbf{b}}_x = \frac{\mathbf{e}_1}{\|\mathbf{e}_1\|} \text{ and } \tilde{\mathbf{b}}_y = \frac{\mathbf{e}_2}{\|\mathbf{e}_2\|} \qquad (5.2.13)$$

form the pair of surface base vectors, which are orthogonal and aligned with the surface (X and Y vectors in figure 5.4).

Surface size estimation

Afterwards, the dimensions of the surface along the principal components can be determined. This is accomplished by finding the points with the largest distance to the centroid along the principal components:

$$\tilde{\Gamma} = \begin{pmatrix} \max_{\mathbf{p} \in \mathcal{M}} \left((\mathbf{p} - \tilde{\mathbf{c}}) \cdot \tilde{\mathbf{b}}_x \right) - \min_{\mathbf{p} \in \mathcal{M}} \left((\mathbf{p} - \tilde{\mathbf{c}}) \cdot \tilde{\mathbf{b}}_x \right) \\ \max_{\mathbf{p} \in \mathcal{M}} \left((\mathbf{p} - \tilde{\mathbf{c}}) \cdot \tilde{\mathbf{b}}_y \right) - \min_{\mathbf{p} \in \mathcal{M}} \left((\mathbf{p} - \tilde{\mathbf{c}}) \cdot \tilde{\mathbf{b}}_y \right) \end{pmatrix} \qquad (5.2.14)$$

Surface selection

The next step is to prepare the calculations in the satellite body frame. First, the model surface centroid $\mathbf{c}(s)$ must be transformed into the satellite body frame. For the normal vector $\mathbf{n}(s)$ and the two principal components $\mathbf{b}_x(s)$, $\mathbf{b}_y(s)$, only the rotation quaternion of the transformation is applied.

Then, matching surfaces are found by reducing the model surface set $\mathcal{S}$ to the reduced surface set $\mathcal{S}'$, such that

$$\mathcal{S}' = \left\{ s \in \mathcal{S} : \mathbf{c}(s) \cdot \mathbf{n}(s) > 0 \wedge \left\| \Gamma(s) - \tilde{\Gamma} \right\|_1 < \zeta \right\}, \qquad (5.2.15)$$

where $\|\cdot\|_1$ is the *sum norm*, $\Gamma(k)$ is the dimension vector of the surface k (as it was determined in section 3.2.5 on page 88) and ζ is the total maximum of size deviation allowed for two surfaces to match. In the implementation, $\zeta = 0.5$ lead to good results.

The reduced surface set does not contain surfaces, which are on the inside and it does not contain surfaces, which have a large size deviation. For an asymmetric model, $\|\mathcal{S}'\| = 1$. For symmetric models, the correct surface can not be reliably found and must be provided in some way. Multiple pose hypotheses are possible.[2]

Aligning normal vectors and surfaces

Assuming the correct surface $\hat{s}$ has been found, the rotation quaternion is composed from the quaternion that brings the normal axes of the model surface and the measured normal vector into alignment,

$$\mathbf{q}_{\text{align}} = \left((\mathbf{n}(\hat{s}) \times (-\tilde{\mathbf{n}}(\hat{s}))) \cdot \sin\left(\frac{\angle(\mathbf{n}(\hat{s}), \tilde{\mathbf{n}}(\hat{s}))}{2} \right), \cos\left(\frac{\angle(\mathbf{n}(\hat{s}), \tilde{\mathbf{n}}(\hat{s}))}{2} \right) \right), \qquad (5.2.16)$$

---

[2] See section 5.2.5 on page 127.

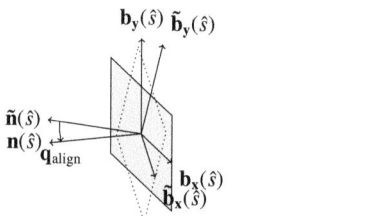 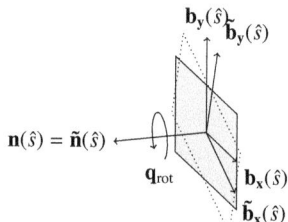

(a) The normal vectors of the measured surface ($\tilde{\mathbf{n}}(\hat{s})$) and the model surface ($\mathbf{n}(\hat{s})$) are brought into alignment.

(b) The principal components are brought into alignment by applying a rotation about the normal vector.

Figure 5.5. Initial attitude quaternion determination. First, the normal vectors of the measured surface and the model surface are aligned using $\mathbf{q}_{\text{align}}$. Then, the model surface is rotated until its principal components match the ones of the point cloud using $\mathbf{q}_{\text{rot}}$.

and the quaternion

$$\mathbf{q}_{\text{rot}} = \left(\tilde{\mathbf{n}}(\hat{s}) \cdot \sin\left(\frac{\angle\left(\mathbf{b}_x(\hat{s}), \tilde{\mathbf{b}}_x(\hat{s})\right)}{2} \right), \cos\left(\frac{\angle\left(\mathbf{b}_x(\hat{s}), \tilde{\mathbf{b}}_x(\hat{s})\right)}{2} \right) \right) \quad (5.2.17)$$

which rotates the model surface, so that the principal components are aligned as well. The resulting quaternion consequently becomes

$$\mathbf{q}_{\text{initial}} = \mathbf{q}_{\text{rot}} \odot \mathbf{q}_{\text{align}}. \quad (5.2.18)$$

Now, the rotation of the target object has been completely estimated. What remains to be done is the estimation of the translational components.

As the model has now been rotated to match the orientation of the measured surface, the new centroid vector becomes

$$\mathbf{c}_r = \mathbf{q}_{\text{initial}}(\mathbf{c}), \quad (5.2.19)$$

where $\mathbf{c}$ is the centroid of the non-transformed model surface. Then, the only difference remaining is the translational offset, what can be determined by simply evaluating the difference between the two centroids,

$$\mathbf{t}_{\text{initial}} = \tilde{\mathbf{c}} - \mathbf{c}_r. \quad (5.2.20)$$

Finally, the vector ($\mathbf{q}_{\text{initial}}, \mathbf{t}_{\text{initial}}$) provides the initial pose estimate. It may be imprecise to a certain degree and it will probably lack precision especially for the translational part. However, it is sufficient to initialize the tracker.

5.2.5 Object symmetry related pose ambiguity

As stated earlier, for highly symmetric objects, the method just presented can not reliably determine all six degrees of freedom. The problem solution envisaged here is twofold. First, when the tracker diverges and the initializer is used to re-initialize the tracker, the attitude quaternion may be checked to minimize a (vector) distance metric over the set of model surfaces

$$\xi = \arg\min_{s \in S} \|\mathbf{q}_s - \mathbf{q}_{\text{last}}\|^2. \tag{5.2.21}$$

In the case where the tracker must be initialized for the first time, hints may be given to the tracker. Hints are small pieces of information which allow the initializer to narrow down the set of possible solutions (poses) until only one pose remains.

By default, the rotation quaternion $\mathbf{q}_{\text{rot}}$ rotates the surfaces in such a way that the principal components are aligned and then, the image is probed with all relevant surfaces with different rotations, where the rotation about the normal vector ($\mathbf{q}_{\text{rot}}$) is changed in steps of 45°. The initialization problem then can be stated as

$$\mathbf{p} = \arg\max_{\mathbf{u} \in \mathcal{U}} \sum_{(x,y) \in \Omega_{\text{PMD}}} \begin{cases} 1 & \mathbf{X}_{\mathbf{u},\text{model}}[y, x] + \mathbf{X}_{\text{measurement}}[y, x] = 2, \\ 0 & \text{otherwise,} \end{cases} \tag{5.2.22}$$

where $\mathcal{U}$ is the set of probed poses (determined as described in the above paragraph) and the two matrices $\mathbf{X}$ are the target affinity matrices – one for the model (which depends on the probed pose $\mathbf{u}$) and one for the point cloud determined from the sensor data. For symmetric targets, the number of probed poses can be reduced by hints:

- When a *surface rotation hint* is given, $\mathbf{q}_{\text{rot}}$ is set to a fixed value and no other configurations of the surface will be tested.

- When an *attitude hint* is given, the attitude quaternion from the hint is used and no matching checks are performed at all. The initializer then only determines the translational components of the complete target pose.

- When one or more *surface exclusion hints* are given, only surfaces not being excluded from the search are probed.

- When a *surface lock hint* is given, only the given surface will be probed. This is particularly useful, when it is known that the target object will be faced with a specific side of its body.

- When a *rotation probe grid hint* is given, the step size for probed orientations (default 45°) can be changed.

Refer to section 8.4 on page 204 for possible additional improvements suggested for future development.

5.3 Summary

In this chapter, the problem of finding the initial relative pose of an object has been discussed. From the review of several publications it became evident that each method depends both on the application and the sensors. Therefore, the same applies to the method presented in the remainder of the chapter.

For minimizing the dependence on environmental lighting, it was decided to use solely the PMD camera for retrieving an initial relative pose estimate. The PMD camera can perform this measurement in complete absence of sunlight, what is not possible with the CCD camera.

The envisaged method is a new approach to the object detection and early pose estimation problem. It uses the amplitude channel of the PMD camera to locate the target object in image-space and then expands a point cloud in 3D space starting from a bootstrap point selected from accumulated amplitude readings. The resulting point cloud provides a centroid, which is used to determine the translational components of the relative pose. A sequential probing of all the model surfaces to the point cloud provides the planar orientation of the matching model surface, and finally, the determination of the principal components of the measured point cloud provides the rotation about the normal vector, what completes the initial relative pose estimate.

In the next chapter, the relative pose tracker is presented, which uses both measurements from the CCD camera and the PMD camera (in the fashion already outlined in section 3.1.5 on page 76) to refine and update the pose estimate during a rendezvous maneuver.

6 High accuracy fused target tracking

Once a rough pose estimate is available, it is necessary to update the pose of the target object, as it moves and rotates over time, while requiring the lowest possible amount of computational power in the process. In this chapter, a novel approach for tracking the target (i.e., iteratively updating the pose with high accuracy) is presented. Also, relevant literature is reviewed. The tracking method presented in the following uses both the CCD- and the PMD sensor and gains its accuracy from the advantages of the sensors, which are neatly fused on a rather high, near-application level.

6.1 Related work

To start with the literature review, a few representative data fusion approaches of PMD- and monocular cameras are presented. Some of the papers use RGB cameras what means even color information is available in the resulting image. Comparable data fusion methods regarding a PMD sensor and a CCD sensor can be grouped together into the high-resolution, low-level interpolation category, where a large number of image pixels is interpolated by different means due to the low lateral resolution of the PMD camera. The result is always a high-resolution image enhanced with distance information.

Figure 6.1 provides a graphical distinction of the different types of sensor data fusion approaches. The difference is the level where the data fusion takes place.

In a more general context, high-level data fusion as proposed within this thesis is also often performed by using different variants of the Kalman filter [36, 67, 99] or the particle filter [117], but these approaches require a complete pose estimate from each sensor. In contrast, here it is attempted to fuse the sensor data based on

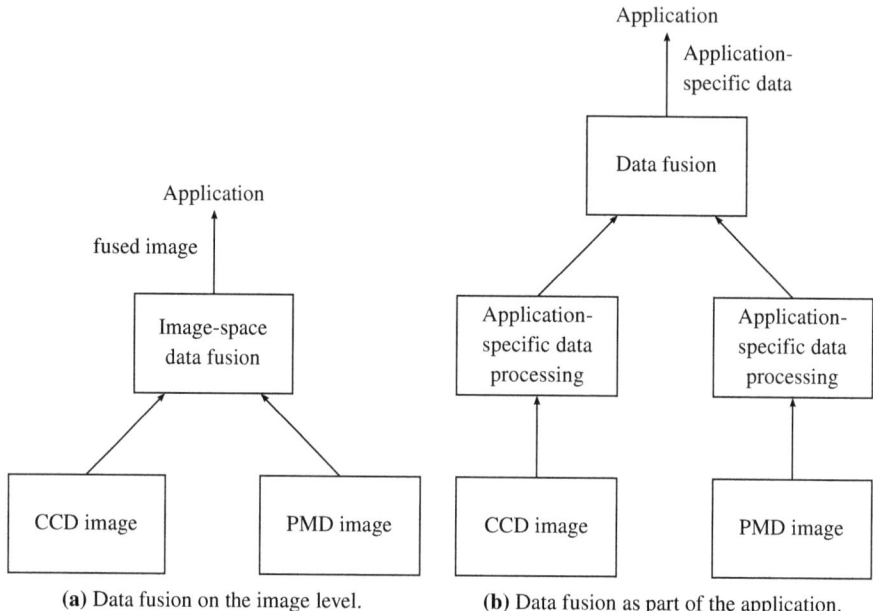

Figure 6.1. Different approaches for sensor data fusion. The left image shows the common way of first fusing the image data of both sensors on the image-level, before any application-specific processing takes place. On the other hand, on the right, a high-level data fusion approach is shown, where the data fusion is itself part of the application. The approach taken with this thesis is, in its structure, similar to the right figure.

distributed measurements of position and orientation deviations in order to obtain a complete pose estimate.

6.1.1 Scene acquisition

Fusing PMD measurements with monochromatic CCD image data is found rarely [114] compared to approaches using RGB cameras [60, 64, 65, 87, 89, 116]. One method of fusing the data relies on the orthophoto approach [94]. This method is used to map PMD pixels into the CCD camera frame. Basically, it works by recalculating the PMD image in such a way that the 3D information becomes a height map of a 2D image plane. After this step, the resulting height map can be scaled and applied to the high-resolution CCD image.

There are also approaches interesting particularly from the hardware side [106], when using an optical splitter in order to combine both sensors for using single optics. The large advantage of this is, that a lot of calibration work is not necessary – both sensors basically use the same coordinate frame. The drawback is the lowered sensitivity of both sensors caused by the optical splitter. As only approximately half of the light falls on each of the sensors, especially the range measurement can become difficult for objects with dark surfaces or low reflectivity in general.

The problem of model building using a PMD sensor together with a high-resolution CCD camera can be tackled by first translating the point cloud measured by the PMD camera into the CCD camera coordinate frame. Then, Markov Random Fields [84] may be used to fuse the images [60]. In essence, such methods are based on the assumption that image space color or brightness discontinuities also cause depth discontinuities. The method can also be applied to and was originally presented in conjunction with a laser scanner [29]. In the end, this provides a very smooth, high-resolution image enhanced with depth information, while conserving edge information. A drawback is that this algorithm is sensitive to strong noise and highly textured surfaces, for example those produced by MLI shielding.

Fusing stereo vision data with PMD data has also been investigated [52, 76, 98], as each sensor may compensate for the deficiencies of the other one. In more detail, the PMD camera provides good measurements especially for unstructured surfaces, where stereo vision fails. On the other hand, for a very much detailed structure, stereo vision provides a good distance estimate, since there is a high number of clues for estimating the disparity. For fusing the data, the PMD camera is used as a rough initializer for the stereo matching algorithms. In a later stage, confidence intervals are used to choose whether to trust the PMD measurement or the stereo matching result. In a different approach, fusing both stereo vision and PMD distance data into a single image is even proposed without extrinsic calibration [54] using graph cuts [51].

6.1.2 Scene segmentation and object recognition

There are also attempts of RGB camera/PMD camera data fusion where object recognition is the high-level goal [87]. Here, instead of recalculating the PMD camera image, one can rely on projective texture maps [65, 129], where the CCD camera image is projected onto the 3D shape provided by the PMD camera. For speeding up the calculations, the data fusion can be performed on a graphics processing unit (GPU). The advantage is the faster processing speed. The drawback is the dependency on the hardware. Unfortunately, especially in space, highly parallelized and powerful processing units – as can be found on today's graphics cards – are not available.

In a follow-up paper [89], the method was improved by refining the distance estimate near and around edges by a more suitable interpolation [135].

Image segmentation algorithms can profit from the depth measurements of a PMD camera [123]. In this particular application, an electronic chalk board and a teacher in front is being used. The PMD camera provides depth measurements, which are transformed into the CCD coordinate frame. Then, a depth thresholding is made after an erode step in order to remove border effects. Color samples of both depth areas are then used to initialize and update the segmentation provided by the SIOX algorithm (see [44] and [43] for details).

Objects can also be identified using groups of features [124]. Here, such groups are mapped on object surfaces and a neighborhood relation between surfaces allows complete description of the object, what defines the model. When capturing an object from different viewpoints, background features become unstable immediately. This provides a way of filtering such features, as they do not form a group of stable features.

Purely PMD-sensor based approaches are also possible [152] but depend on the accuracy requirements, as high-resolution CCD sensors are still superior to localizing certain image features [95].

6.1.3 Pose estimation

The pose estimation problem is a common topic in all image- and sensor data processing articles. One possible approach is to fit a plane to carefully selected key points [110, 112], where edges are detected in the image using a Canny edge detector. Then, on two outer edges, three points are selected. By intersection lines perpendicular to the edges, nine more points located on the surface of the object are retrieved. By finally reading out the distance value and fitting a plane to the resulting points, the pose is recovered. This approach does not rely on a previously defined model or features, but it can directly estimate the pose simply and fast. The drawback of this approach is the low number of pixels used for the actual measurement, what makes the method prone to noise and outliers.

A different approach, which is more close to the method presented in this thesis is the combination of least-squares plane fitting using the PMD point cloud and the Hough transform [35] to determine the orientation of edges in the amplitude channel of the camera [111], resulting in significantly improved precision.

When using stereo vision, the pose of objects can also be estimated by tracking the lines of a known model and using the 3D information along with the 2D edges [69]. Here, for higher accuracy, the iterative least-squares pose refinement algorithm [34] was used: Instead of using an algorithm such as RANSAC [38] to filter outliers, a weighting function is introduced, which depends on a spatial distance parameter changing with each iteration. This results in very good robustness, especially in case of partial occlusion. Using the point cloud and fitting a model to it using a least squares method is also possible [59].

Finally, the author has already implemented a feature-based pose estimation algorithm for Rendezvous and Docking scenarios in the past [143]. To introduce the main idea briefly, while feature extractors such as SURF [11] and SIFT [93] are commonly used in 2D images captured by CCD sensors, in the paper it was attempted to create a novel feature extractor capable of extracting 3D features. This works by using the depth information of the camera to retrieve surfaces and fit extraction planes to these surfaces. Once found, local reflection properties of the object are extracted in a way tolerant to arbitrary rotations. However, the weakness of this approach are surfaces with a lot of structure in terms of physical shape.

To summarize, what can be seen from the reviewed literature is that data fusion often happens on the image level and the accuracy of single-sensor approaches is often very limited. There is a large number of data fusion methods with different advantages and drawbacks, but there seems to be a lack of inclusion of application requirements. Granted, for mobile robot localization and mapping, there is not much room for specializing, however, for estimating the pose of an object, a lot more can be done to improve the overall result.

6.2 Geometric feature estimation

Beginning with this section, the tracking algorithm is presented. In this case, tracking means to refine a given pose estimate and update it on a frame-by-frame basis. The initial pose estimate may be given by the user, an initializing algorithm (such as the one presented in chapter 5 starting on page 115), a navigation filter, or some other source.

6.2.1 Dominant surface

The first step in the on-line processing part is to retrieve the point cloud from the PMD camera. Let $\mathbf{D}$ be the matrix of distance measurements of the PMD camera and let $\mathbf{r_{x,y}}$ be the direction vector of the image pixel coordinate (x, y) (calculated in the off-line preprocessing stage – see section 3.3 on page 88), then the point cloud becomes

$$\mathcal{M} = \bigcup_{(x,y) \in \Omega_{PMD}} \mathbf{D}[y, x] \cdot \mathbf{r_{x,y}}. \qquad (6.2.1)$$

The next step is to determine the *dominant surface* of the model. This is accomplished by algorithm 6.1.

Definition 6.1 (Dominant surface) *Given the optical axis vector $\mathbf{z}$ of the sensor and the set of transformed model surfaces $\mathcal{S}$, the dominant surface is the surface $s \in \mathcal{S}$ which maximizes $\prod_k \Gamma_k(s)$ (the total area) and minimizes $\angle(-\mathbf{z}, \mathbf{n}(s))$ (the angle between the inversed optical axis vector and the surface normal vector).*

6.2.2 Surface affinity matrix

After the dominant surface is known, the surface affinity matrix $\mathbf{X}$ can be retrieved, which contains the information of whether a pixel is part of the target object or not.

Definition 6.2 (Surface affinity matrix) *A matrix $\mathbf{X}$ which is of the same size as the distance matrix $\mathbf{D}$, each element corresponding to an image pixel, is called surface affinity matrix, when for each element*

$$\mathbf{X}[y, x] = \begin{cases} 1 & \textit{The projection of the surface contains } (x, y) \\ 0 & \textit{otherwise.} \end{cases}$$

Algorithm 6.1. Determination of the dominant surface.

Input	: Optical axis vector $\mathbf{z}$	
Input	: Set of transformed model surfaces S	
Input	: Current pose rotation quaternion $\mathbf{q_p}$	
Output	: Dominant surface $\hat{s}$	
1	$b_{best} \leftarrow \pi$	// best angular deviation
2	$a_{best} \leftarrow 0$	// largest area
3	$\hat{s} = \epsilon$	
4	**forall the** $s \in S$ **do**	
5	$\quad \mathbf{v} \leftarrow \mathbf{q_p} \odot (\mathbf{n}(s), 0) \odot \overline{\mathbf{q_p}}$	// apply current pose (rotation only)
6	$\quad$ **if** $(\angle(\mathbf{v'}, -\mathbf{z}) \leq b_{best})$ **then**	// Angular improvement?
7	$\quad\quad$ **if** $(a(s) > a_{best})$ **then**	// Area improvement?
8	$\quad\quad\quad b_{best} \leftarrow \angle(\mathbf{v'}, -\mathbf{z})$	
9	$\quad\quad\quad a_{best} \leftarrow a(s)$	
10	$\quad\quad\quad \hat{s} \leftarrow s$	

The matrix is constructed by first initializing all elements to zero. Then, the scanline algorithm[1] changes elements to 1 where a triangle is drawn. The reasoning for choosing this particular method will be given later in section 6.4.1 (page 143). The entire algorithm is too long to put it here, therefore, a summary is given.

For three given triangle vertices, the algorithm starts at the upper left point and from there moves to the right, until the right border of the triangle is hit. Then, it moves one line down and repeats. This happens until the entire triangle has been drawn. A visualization is provided in figure 6.2. The large size of the algorithm results from the number of different cases, which must be handled depending on the orientation and shape of the triangle.

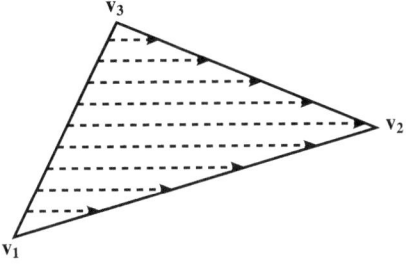

Figure 6.2. Principle of the Scanline algorithm. Triangles are drawn by traversing each line of pixels from the upper left to the bottom right point.

---

[1] Despite the name of the algorithm, this has nothing to do with the scanlines used in later stages. While this algorithm is dedicated to drawing triangles, the scanlines used later are needed for finding the position of an edge.

Before the triangles can be drawn, however, their 2D image space vertex coordinates must be known. First, for all triangle vertices $x \in \mathcal{V}(t)$ and for all triangles $t \in \mathcal{T}(\hat{s})$ (where $\hat{s}$ is the currently tracked surface), the current pose is applied:

$$\mathbf{x_p} = \mathbf{q_p} \odot (\mathbf{x}, 0) \odot \overline{\mathbf{q_p}} + (\mathbf{t_p}, 0), \tag{6.2.2}$$

where $\mathbf{q_p}$ is the rotation quaternion of the current pose estimate and $\mathbf{t_p}$ is the translational part of the current pose estimate. Next, the individual sensor reference frame must be applied; due to its definition, however, it must be applied inversed, such that

$$\mathbf{x_s} = \overline{\mathbf{q_s}} \odot \mathbf{x_p} \odot \mathbf{q_s} - (\mathbf{t_s}, 0), \tag{6.2.3}$$

where $\mathbf{q_s}$ and $\mathbf{t_s}$ are the rotation quaternion and translation vector of the sensor, respectively. Next, the resulting 3D vertex coordinate, which is now relative to the sensor coordinate frame, is projected into 2D image space. First, the normalized coordinates

$$\mathbf{x_u} = \begin{pmatrix} -x_{s_1}/x_{s_3} \\ -x_{s_2}/x_{s_3} \end{pmatrix} \tag{6.2.4}$$

are retrieved. Then, optical distortion must be respected. Here, the complete distortion model is applied (meaning tangential and radial distortion). The formulas were taken from the CalTech Camera Calibration Toolbox. The resulting distorted pixel coordinate then is

$$\mathbf{x_d} = \begin{pmatrix} \left(1 + k_0 \cdot r + k_1 \cdot r^2\right) \cdot x_{u_0} + 2 \cdot k_2 \cdot x_{u_0} \cdot x_{u_1} + k_3 \cdot \left(r + 2 \cdot x_{u_0}^2\right) \\ \left(1 + k_0 \cdot r + k_1 \cdot r^2\right) \cdot x_{u_1} + 2 \cdot k_3 \cdot x_{u_0} \cdot x_{u_1} + k_3 \cdot \left(r + 2 \cdot x_{u_1}^2\right) \end{pmatrix}, \tag{6.2.5}$$

where $r = \|\mathbf{x_u}\|^2$, the squared norm of the undistorted pixel coordinate vector $\mathbf{x_u}$. After this step, the vector must be multiplied by the focal length and the principal point must be added as offset in order to obtain image space coordinates, what yields

$$\mathbf{x} = \begin{pmatrix} f_x \cdot x_{d_0} + c_x \\ f_y \cdot x_{d_1} + c_y \end{pmatrix}. \tag{6.2.6}$$

This calculation can now be made for each 3D vertex of the model belonging to the visible surface. Then, so far, the triangles are drawn using the Scanline algorithm and the pixels belonging to the dominant surface are memorized in the surface affinity matrix $\mathbf{X}$. This is done in the image space of the PMD sensor.

6.3 Scanline processing

Equally, after all model lines have been projected into the image space of the CCD sensor, scanlines can be extracted to track the lines in the image. As it was discussed in section 3.1.5 (page 76), here, data processing is limited to estimating residual data for only three degrees of freedom. In the end, the displacement in X- and Y-direction as well as the angular displacement about the roll axis is to be determined.

In order to achieve this, articulated edges of the target object are tracked. For making the tracking process computationally efficient, the number of pixels sampled during this step is minimized by only sampling a particular selection of image pixels. These pixels will have the shape of lines, which will be referred to as *scanlines* in the following.

The idea of using lines in such a way for tracking edges is common practice in machine vision, as this is a problem which has been addressed for a long time [33, 34, 55, 132], [61, pp. 368–369].

Definition 6.3 (Scanline) *Given the set of image pixels $\mathcal{P}$, a scanline is a sequence of pixels $\mathcal{P}' \subset \mathcal{P}$ chosen in such a way that it forms a line crossing an articulated edge of the model clearly visible to the texture segmenter in the CCD image. A natural ordering "<" is defined among the set elements of $\mathcal{P}'$, allowing the line to be traversed.*

6.3.1 Scanline parameters

Scanlines are used to gather information about the change in position and orientation of an articulated edge visible in the CCD image stream. To accomplish that, scanlines are best placed perpendicularly along the last known line of the model. This line is obtained by projecting all lines $l \in \mathcal{L}(\hat{s})$ belonging to the currently tracked surface $\hat{s}$ of the model using the last known model pose $\mathbf{p}$, into the CCD image plane. In the following, $\sum_l$ will denote the summation over all $l \in \mathcal{L}(\hat{s})$. For the remainder of this section, please have a look at figure 6.3 for help in understanding the process.

Each one of these lines has a starting point $\mathbf{l_s}$ and an endpoint $\mathbf{l_e}$ in CCD image space. The vector describing the difference is thus $\mathbf{m}_l = \mathbf{l_e} - \mathbf{l_s}$. Then, for each line

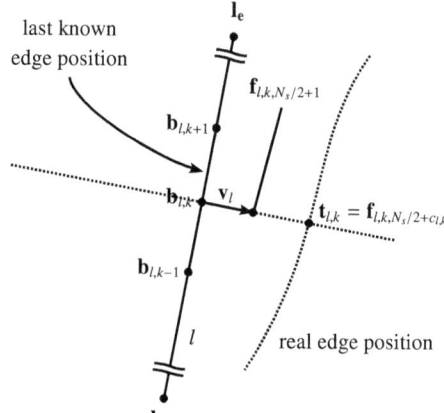

Figure 6.3. Visualization of a scanline and edge finding. A scanline is used to perpendicularly search for an edge in the image, relative to a last known line.

$l \in \mathcal{L}(\hat{s})$, a *scanline spawn point* $\mathbf{b}_{l,k}$ is computed by traversing the line, such that

$$\mathbf{b}_{l,k} = \mathbf{l}_s + \frac{k}{N_l}\mathbf{m}_l, \tag{6.3.1}$$

where k is a natural number which increases from 0 up to $N_l - 1$. N_l is the *number of planned scanlines* for the model line l. It is determined by the length of the line $\|\mathbf{m}_l\|$, such that

$$N_l = N_s \cdot \frac{\|\mathbf{m}_l\|}{\sum_{x \in \mathcal{L}(\hat{s})} \|\mathbf{m}_x\|}, \tag{6.3.2}$$

the ratio between the length of the line l and the total length of all lines $x \in \mathcal{L}(\hat{s})$. It further depends on N_s, the total number of scanlines to use for tracking the entire surface $\hat{s}$, what was found to provide good results with $N_s = 80$.

Now, after the scanline spawn point has been determined, the scanline is placed perpendicular to the model line, with its central point coinciding with the scanline spawn point. For that, the direction vector

$$\mathbf{v}_l = \frac{\mathbf{m}_l}{\|\mathbf{m}_l\|} \times \begin{pmatrix} 0 & 1 \\ -1 & 0 \end{pmatrix} \tag{6.3.3}$$

is created. After that, the pixels are sampled from image space, according to the position vector

$$\mathbf{f}_{l,k,x} = \mathbf{b}_{l,k} + \left(\frac{x}{\gamma/2} - \gamma\right) \cdot \mathbf{v}_l, \tag{6.3.4}$$

where x is a natural number increasing from 0 to $\gamma - 1$ and $\gamma = 50$ is the length of the scanline. Finally,

$$\mathbf{s}_{l,k} = \mathbf{I}\left[\mathbf{f}_{l,k,x}\right]\Big|_{x \in [0, \gamma - 1] \cap \mathbb{N}_0} \tag{6.3.5}$$

contains the image space samples of the pixels targeted by the k-th scanline on the projected model line l.

6.3.2 Texture change point

Assuming the image space samples have been successfully obtained in $\mathbf{s}_{l,k}$, the texture change point can be determined. Please refer to the introductory section 2.6.1 on page 66 for more information about the following texture segmentation algorithm. At this point, only the actual implementation will be discussed.

First of all, the sampled pixel values may be limited in dynamic range. For example, if a dark part of the image is sampled, the values in the scanline vector $\mathbf{s}_{l,k}$ will have a very limited value range. Since the brightness values must be quantized (due to the very limited length of the scanline), a normalization is required, such that

$$\tilde{\mathbf{s}}_{l,k} = \frac{\mathbf{s}_{l,k}}{\max \mathbf{s}_{l,k}}. \tag{6.3.6}$$

Next, the brightness values in the normalized scanline vector $\tilde{\mathbf{s}}_{l,k}$ are quantized into bins. This is required since the ratio of the length of the scanline and the bit depth of the image sensor is too large. Note that quantization is not necessary when the scanline length is of the same magnitude (or even larger) than the number of possible brightness values.

Here, the normalized and quantized brightness values are denoted by

$$\overline{\mathbf{s}_{l,k}} = \left\lfloor \tilde{\mathbf{s}}_{l,k} \cdot \left(2^{\tilde{b}} - 1\right) \right\rfloor, \tag{6.3.7}$$

where $\tilde{b}$ is the reduced bit depth, such that 2^{N_q} is the number of quantization bins. $\overline{\mathbf{s}_{l,k}}$ now only contains numbers in $[0, (2^{N_q} - 1)]$.

Afterwards, the texture change point estimation can be performed by maximizing the product of the probability term

$$p_{l,k}(u,v) = \prod_{z=u}^{v} \frac{\sum_{h=u}^{z} 1 + \begin{cases} 1, & \overline{s_{l,k}}(z) = \overline{s_{l,k}}(h) \\ 0, & \text{otherwise} \end{cases}}{2N_q - 1 + (z - u + 1)} \qquad (6.3.8)$$

of the two texture-generating processes T_1 and T_2, such that the texture change point becomes

$$c_{l,k} = \underset{c \in [\alpha, \gamma-1-\alpha]}{\arg\max} \; p_{l,k}(0, c) \cdot p_{l,k}(c+1, \gamma-1), \qquad (6.3.9)$$

where α is the minimum number of samples considered for a texture generating process, $p_{l,k}(0, c)$ is the probability of the sequence $\{\overline{s_{l,k}}(0), \ldots, \overline{s_{l,k}}(c)\}$ belonging to T_1 and $p_{l,k}(c+1, \gamma-1)$ for T_2, respectively. In the current implementation, $\alpha = 4$. The idea is that when sampling the two texture patterns with the correct texture change point, their probability will be maximal and sampling at a wrongly chosen texture change point will only decrease the product of their probabilites. Consequently, the goal is to maximize the product.

Now, after the index of the pixel on the scanline has been determined, it is straightforward to obtain the actual texture change point

$$\mathbf{t}_{l,k} = \mathbf{b}_{l,k} + \left(\frac{c_{l,k}}{\gamma/2} - \gamma\right) \cdot \mathbf{v}_l, \qquad (6.3.10)$$

which is just equation 6.3.4, but with x replaced by $c_{l,k}$.

Before the texture change points can be used, the RANSAC algorithm (see [38] for details) is applied to the points in order to remove outliers (remember that the sequence of texture change points forms a line). In the following, only texture change points remaining after this step are considered. Consequently, $\tilde{N}_s$ will denote the total *reduced* number of scanlines (or resulting texture change points, for that matter) of the surface, which are left after the RANSAC filtering. $\tilde{N}_l$ will denote the reduced number of scanlines on the line l, respectively.

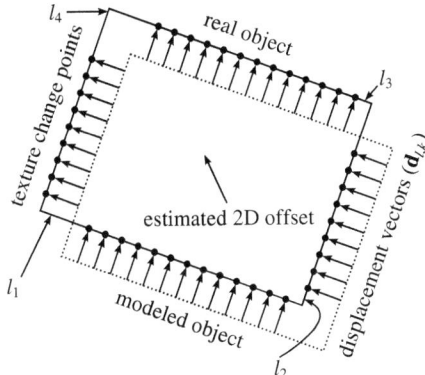

Figure 6.4. Visualization of 2D image space deviation components. A resulting displacement vector is obtained by arithmetic averaging all displacement vectors from all scanlines.

6.3.3 2D image space deviation components

For computing the two first translational deviation components, the distance vector between each texture change point $\mathbf{t}_{l,k}$ and the model line l

$$\mathbf{d}_{l,k} = \mathbf{t}_{l,k} - \mathbf{b}_{l,k} \qquad (6.3.11)$$

is computed. Then, the sum of these vectors provides the deviation components

$$\begin{pmatrix} \Delta x \\ \Delta y \end{pmatrix} = \frac{1}{\tilde{N}_l} \sum_l \sum_{k=0}^{\tilde{N}_l - 1} \mathbf{d}_{l,k}. \qquad (6.3.12)$$

This is shown in figure 6.4. Afterwards, the roll angle residual can be obtained by computing the mean angular deviation of all lines l of the tracked surface $\hat{s}$. Before this can be done, the position deviation must be considered. To regain independence, the centroid of all texture change points

$$\mathbf{g}_l = \frac{1}{\tilde{N}_l} \cdot \sum_{k=0}^{\tilde{N}_l - 1} \mathbf{t}_{l,k} \qquad (6.3.13)$$

is determined, where $\tilde{N}_l$ is the reduced number of scanlines of the line l again.

In the next step, vectors pointing from the texture change point centroid to each particular texture change point are constructed,

$$\mathbf{r}_{l,k} = \mathbf{t}_{l,k} - \mathbf{g}_l. \qquad (6.3.14)$$

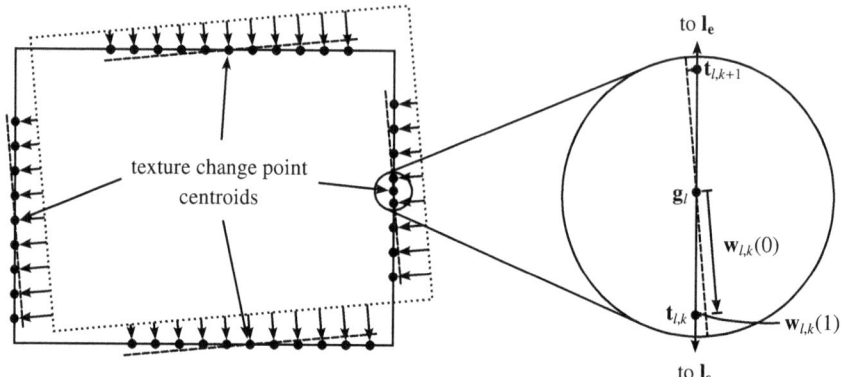

Figure 6.5. Roll angle residual visualization. The residual can be measured once the texture change point centroids have been obtained.

Now the resulting vectors $\mathbf{r}_{l,k}$ are independent of the current model position. In order to calculate the angular deviation, these vectors must be expressed in a different coordinate frame aligned with the model line l. This is done by projecting the vectors onto the base vectors of the targeted coordinate frame, such that

$$\mathbf{w}_{l,k} = \begin{pmatrix} \mathbf{r}_{l,k} \cdot (\mathbf{m}_l / \|\mathbf{m}_l\|) \\ \mathbf{r}_{l,k} \cdot \mathbf{v}_l \end{pmatrix}, \qquad (6.3.15)$$

where $\mathbf{m}_l / \|\mathbf{m}_l\|$ is the normalized direction vector of the line l (note that in the previous definition of $\mathbf{m}_l$, it was intentionally not normalized) and $\mathbf{v}_l$ is the scanline direction vector (which was normalized by definition).

The angular displacement can now be retrieved as the angle between the zero vector (first base vector in the right-handed, 2D coordinate frame) and $\mathbf{w}_{l,k}$ (figure 6.5). As a consequence, it follows that

$$\Delta\gamma = \frac{1}{\#\mathcal{L}(\hat{s})} \sum_{l \in \mathcal{L}(\hat{s})} \sum_{k=0}^{\tilde{N}_l - 1} \arctan 2\left(\mathbf{w}_{l,k}(1), \mathbf{w}_{l,k}(0)\right), \qquad (6.3.16)$$

what completes the processing of the CCD sensor data.

6.4 Point cloud processing

After the CCD sensor data has been processed, the data of the PMD sensor must be prepared and then processed. In section 6.2 (page 134), the processing of model data along with coordinate frame transformations and optical distortion accommodation has already been described. In the following, the surface affinity matrix is retrieved from the model and then border effects are addressed.

6.4.1 Surface affinity determination

When processing images from the PMD sensor, the object must be separated from the background and even more important, surfaces currently being tracked must be separated from remaining parts of the object. While approaches of using image data for performing this separation are possible, there are often drawbacks caused by limited accuracy, outliers and singularities. Therefore, the surface affinity is determined by using the model information.

When the current (approximate) pose is known, the model along with all of its surfaces can be projected into the PMD camera frame and then pixels belonging to an object surface can be marked. This is the high-level idea. On a lower level, there are multiple methods for marking the pixels, assuming the model defines the surfaces as sets of triangles (of which both have been tested):

- **Raytracing.** Here, for each pixel of the camera sensor, a virtual ray is cast into the scene and every triangle is checked for a collision with the ray. The collision closest to the camera defines the distance and surface of that pixel. This method is very accurate and provides barycentric coordinates without much additional effort. However, it has the drawback of being too slow.

- **Scanline algorithm.** The triangles are first projected into 2D image space. Then, each triangle is processed individually in the following way: Start at the leftmost, uppermost pixel of the triangle (this will be one of the first vertices after sorting them using their Y coordinate). Then, mark pixels, row by row, from left to right, by walking along the edges of the triangle as borders. This method is well known in computer graphics and utmost fast. The drawback is that barycentric coordinates are not retrieved – additional effort must be put in, to retrieve them.

Due to the speed of the Scanline method and the fact that barycentric coordinates are not important for this application, it is the method of choice. It has already been introduced in section 6.2.2 (page 134). After the projection of the dominant surface of the model, a rough surface affinity matrix is available.

6.4.2 Surface affinity matrix refinement

Since the projection used the pose estimate of the previous frame, there may be a discrepancy between the model and the real object, which will be addressed in the following. Especially pixels near the border of the dominant surface can be assigned to belong to the target instead of the background and vice versa.

For this purpose, a planar RANSAC algorithm is now applied to the model surface described by the respective surface affinity matrix $\mathbf{X}$ in conjunction with the distance matrix $\mathbf{D}$. For better stability, the choice of pixels is straitened by the constraint that for the three points $\mathbf{p}_1, \mathbf{p}_2, \mathbf{p}_3$,

$$\|\mathbf{p}_1 - \mathbf{p}_2\| > 0.5 \wedge \|\mathbf{p}_2 - \mathbf{p}_3\| > 0.5 \wedge \|\mathbf{p}_3 - \mathbf{p}_1\| > 0.5 \qquad (6.4.1)$$

must hold at all times. By enforcing that the three points, which serve as a definition for the RANSAC testing surface, are spatially distributed, a lot of processing time is saved. In the end, after the RANSAC algorithm has finished, the surface affinity matrix $\mathbf{X}$ will be updated accordingly, such that erroneous pixels (too much noise, border effects, sensor saturation, etc.) are removed.

After the surface affinity has been refined, the three object/ model pose deviation components can be calculated, which provide the foundation for the numeric optimization and thus, finding the updated pose estimate.

6.4.3 Deviation component determination

First the surface centroid must be found. Assuming $\mathcal{P}$ is the set of points belonging to the surface $\hat{s}$ (points selected from the sensor point cloud), this becomes

$$\tilde{\mathbf{c}}(\hat{s}) = \frac{1}{\|\mathcal{P}\|} \cdot \sum_{\mathbf{x} \in \mathcal{P}} \mathbf{x}. \qquad (6.4.2)$$

Next, the normal vector of the surface is retrieved by evaluating

$$\tilde{\mathbf{n}}'(\hat{s}) = \left(\sum_{\mathbf{x} \in \mathcal{P}} (\mathbf{x} - \tilde{\mathbf{c}}(\hat{s})) \cdot \frac{x_1 - \tilde{c}_x}{\|\mathbf{x} - \tilde{\mathbf{c}}(\hat{s})\|} \right) \times \left(\sum_{\mathbf{x} \in \mathcal{P}} (\mathbf{x} - \tilde{\mathbf{c}}(\hat{s})) \cdot \frac{x_2 - \tilde{c}_y}{\|\mathbf{x} - \tilde{\mathbf{c}}(\hat{s})\|} \right). \quad (6.4.3)$$

Note that $(x_1 - \tilde{c}_x) / \|\mathbf{x} - \tilde{\mathbf{c}}(\hat{s})\|$ is essentially the cosine of the angle between the difference vector $\mathbf{x} - \tilde{\mathbf{c}}(\hat{s})$ and the X axis, while $(x_2 - \tilde{c}_y)/ \|\mathbf{x} - \tilde{\mathbf{c}}(\hat{s})\|$ is the sine, respectively. Finally, the normal vector is retrieved after the normalization

$$\tilde{\mathbf{n}}(\hat{s}) = \frac{\tilde{\mathbf{n}}'(\hat{s})}{\|\tilde{\mathbf{n}}'(\hat{s})\|}. \quad (6.4.4)$$

Then, the same vector can be used to determine the angular displacements for the pitch angle (see also figure 3.1 on page 76),

$$\Delta\alpha = \arctan2\left(\tilde{n}_z(\hat{s}), \tilde{n}_x(\hat{s})\right) - \arctan2\left(n_z(\hat{s}), n_x(\hat{s})\right), \quad (6.4.5)$$

and the yaw angle,

$$\Delta\beta = \arctan2\left(\tilde{n}_z(\hat{s}), \tilde{n}_y(\hat{s})\right) - \arctan2\left(n_z(\hat{s}), n_y(\hat{s})\right). \quad (6.4.6)$$

In both cases, $\tilde{\mathbf{n}}(\hat{s})$ refers to the *measured* normal vector of the surface $\hat{s}$ and $\mathbf{n}(\hat{s})$ refers to the normal vector of the corresponding model surface. The additional indices denote the vector components.

The distance deviation is obtained by determining the distance vector between a point on the model surface ($\mathbf{p}_{\text{Ref}}(\hat{s})$) and the measured point cloud centroid, then projecting this vector onto the point cloud normal vector:

$$\Delta z = (\tilde{\mathbf{c}}(\hat{s}) - \mathbf{p}_{\text{Ref}}(\hat{s})) \cdot \tilde{\mathbf{n}}(\hat{s}). \quad (6.4.7)$$

This completes the calculation of the last three deviation components. By numerically minimizing these quantities, the pose estimate can be obtained.

6.5 Pose estimation

The pose estimation is mainly inspired by the individual sensor characteristics worked out in table 3.1 (page 78). It is an attempt to maximize the accuracy of the pose estimate by using each sensor's strengths.

6.5.1 Pose representation

The pose of the object is a union of two different operations: Translation and rotation. While the translation can be simply expressed as a vector in 3D space, the rotation can be represented by numerous means, such as Euler angles, a rotation matrix or a quaternion. Rotation matrices are a good choice in situations where a large number of vectors must be rotated. For complete transformations, by using homogeneous coordinates, the translational part and the rotational part can be combined into one single matrix. Standard methods for inverting a matrix can be used to retrieve the inverse transformation. The problem with that is however, that matrix inversions are in general non-trivial and can fail entirely for ill-conditioned matrices.

Euler angles are a very common representation of linked rotations. The main advantage is the easy to read representation for human beings. The drawbacks are the gimbal lock, a singularity where exchanging angles in a certain way will not alter the rotation result and second, the order in which the rotations about the coordinate frame axes are performed, is important and there is, in general, no universally accepted convention for that.

Finally, the quaternion is a computationally very efficient representation of rotations. There are no problems with singularities. This suggests that quaternions are simply the best representation of rotations for technical applications. This is why a quaternion has been chosen to represent the orientation.

Definition 6.4 (Pose vector) *A pose vector is a 7-element vector, composed of a 4-element rotation quaternion* $\mathbf{q} = (q_x, q_y, q_z, q_w)$ *and a 3-element translation vector* $\mathbf{t} = (t_x, t_y, t_z)$ *such that the pose vector is* $\mathbf{p} = (q_x, q_y, q_z, q_w, t_x, t_y, t_z)$.

6.5.2 Pose vector optimization

The algorithm design is a tracker architecture. The idea is to have an approximate pose estimate and refine this estimate with each pair of images retrieved from the sensors. As the pose estimation problem is in general non-linear and a closed-form solution is not available, the method relies on numeric optimization.

There are several optimizers known, however the Levenberg-Marquardt algorithm [92, 97] is a very well documented and proven method, which is why it is being used here. It seeks the minimum of an objective function χ, which in this case is defined as

$$\chi : \mathbb{S}^3 \times \mathbb{R}^3 \times \mathbb{R}^{h \times w} \times \mathbb{R}^{v \times u} \times \mathbb{R}^{v \times u} \to \mathbb{R}^7, \qquad (6.5.1)$$

where $\mathbb{S}^3$ is the set of unit vectors in $\mathbb{R}^4$ (3-sphere). This function maps a pose vector $\mathbf{p} \in \mathbb{S}^3 \times \mathbb{R}^3$, along with the brightness image $\mathbf{B} \in \mathbb{R}^{h \times w}$ retrieved from the CCD camera and the distance image $\mathbf{D} \in \mathbb{R}^{v \times u}$ and the amplitude image $\mathbf{A} \in \mathbb{R}^{v \times u}$ retrieved from the PMD camera (which are all considered constant during the optimization) into the residual space $\mathbb{R}^7$. The residual vector is composed of the angular and translational displacements, as well as the deviation of the quaternion norm from the unit norm (1). A solution is found when all elements of the residual vector are zero.

6.5.3 Error metric

What remains to be done is defining an error metric, which can be minimized in the pose space, so that the pose vector is updated with respect to the incoming sensor images. Table 3.1 on page 78 already listed the quantities that are going to be used. Therefore, the residual can be put together as

$$\epsilon_{\mathbf{p}} = (\Delta\alpha, \Delta\beta, \Delta\gamma, \Delta x, \Delta y, \Delta z, k \cdot (1 - \|\mathbf{q}\|))|_{\mathbf{p}}, \qquad (6.5.2)$$

where $\Delta\alpha$ is the pitch angle deviation, $\Delta\beta$ is the yaw angle deviation, $\Delta\gamma$ is the roll angle deviation, $k \cdot (1 - \|\mathbf{q}\|)$ is the amplified unit quaternion norm deviation, and Δx, Δy and Δz are the translational position deviations, respectively. The unit quaternion deviation is introduced as an additional quantity here in order to prevent the optimizer from changing the norm of the quaternion embedded in the pose vector. The amplification factor has been determined experimentally and it was found that a value of $k = 100$ provided stable results.

The residual can then be written as a sum of squares over the above vector elements, or simply the square of the vector norm, such that

$$\chi(\mathbf{p}) = \|\epsilon_{\mathbf{p}}\|^2 \qquad (6.5.3)$$

becomes the objective function.

The individual components are calculated as described in sections 6.3.3 (page 141, for the 2D deviation components) and 6.4.3 (page 144, for the 3D deviation components).

The optimization is carried out using the Levenberg-Marquardt algorithm [92]. Due to the design of the error metric, the norm of the quaternion will not have diverged to a large degree, but for the stability of the algorithm and further processing of the pose vector, the quaternion is normalized after the convergence of the algorithm so it can be guaranteed that it is of norm 1. After that point, the update of the pose vector is complete.

6.6 Summary

In this chapter, a novel method for tracking a target object was presented. The method utilizes measurements from the CCD camera for tracking edges and PMD camera data for tracking surfaces. On a frame-by-frame basis, displacements are computed which are then used to minimize a common pose error residual. The pose estimate is obtained by adapting the model transformation towards the real-world position and orientation. By combining the information of the PMD sensor and the CCD sensor in this common optimization, the advantages of the sensors are properly used and a highly accurate pose estimate is retrieved.

Compared to existing approaches, where either the images of the two sensors are fused on the image level [60, 94] and the application-level data processing runs on the high-resolution depth image created by the data fusion or the data fusion is accomplished by using a stochastic state estimator such as the Kalman filter [36, 67, 99], here, the data processing is limited to relevant parts of the images only. This is important for minimizing the computational need, what is easily achieved with the proposed method.

Before going into the measurements, it is advisable to have a look at the two flow diagrams presented in chapter 1 again. Figure 1.2 (page 32) summarizes the preprocessing part and figure 1.3 (page 33) shows the principal part of the just presented algorithm. The flow diagrams will be useful for interpreting the results.

In the next chapter, tests are envisaged and conducted in order to assess the performance of the algorithm. Another goal will be to find limits of the method just proposed with regard to possible situations encountered in a space environment.

Part III

Measurements, Analysis & Discussion

7 Performance measurements

So far, the architecture of the proposed algorithm was presented in detail. Despite the postulated theoretical advantages, it is of high importance to test the method in a preferably realistic environment (what very often happens not to be available). For this purpose, the EPOS facility already introduced in section 2.1 (page 35) was used. The test campaign was divided into two major parts; first, it should be demonstrated that the algorithm works as intended. This corresponds to a test under optimal conditions. Then, by applying different lighting conditions, the limit of the proposed method will be found, as the optimal operating conditions are no longer present. Moreover, robustness tests such as temporary loss of tracking will also be considered.

7.1 Setup

In the following, the setup of the testing environment is given. All tests performed on the EPOS facility have this common configuration as a starting point. Deviations from the configuration will be noted in the text.

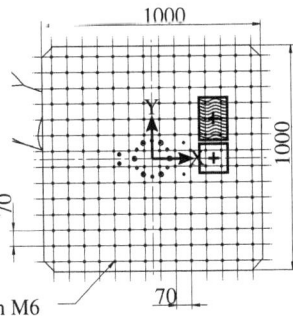

Threadhole pattern M6
pitch: 70 mm
max. torque: 10 Nm

Figure 7.1. CAD drawing of the EPOS mounting plate, including placement of the sensors (clear box: CCD camera, wavy box: PMD camera). Optical axes of the sensors are indicated by crosses. All numbers are in mm. The EPOS tool coordinate frame coincides with the satellite body frame (SBF) used as reference for the pose estimate. *(Mounting plate CAD drawing courtesy of Robo Technology GmbH, used with permission)*

Figure 7.2. Picture of the satellite mockup used for the experiments. The central body of the mockup is 2.3 × 1.8 m in size. It is wrapped in a MLI-like thermal blanket. The apogee nozzle represents the rear of a typical geostationary satellite.

7.1.1 Sensor mounting

The EPOS facility consists of two industrial manipulators, of which both are equipped with a universal mounting plate. To mount the sensors, mounting adapters have been designed which interface to the mounting plate of the EPOS facility and the sensors.

The design of these mounting adapters was lead by the idea of letting the optical axes of the sensors intersect with the center point of the mounting adapter. This way, a precise positioning was made possible without consuming too much space on the EPOS mounting plate.

Figures 7.1 and 2.13 (page 64) show the location of the sensors on the EPOS mounting plate. Although the location of the sensors is known from the mounting process, it is especially the orientation of the sensors, which must be known very precisely. Since the mounting adapters can only provide this precision up to a certain degree despite the efforts of manufacturing them in a local CAM facility, the results of the extrinsic calibration performed in section 2.5 are used as a reference.

7.1.2 Target mockup

A mockup of a typical geostationary satellite has been used as the target object. It is of utmost importance, to have a mockup that is similar to a real satellite as much as possible. Especially the reflectance of infrared light is achieved by using a foil similar to MLI. Figure 7.2 shows an image of the mockup mounted on one of the robots.

The antenna dishes have to be removed during facility operation in order to respect the inertia limits of the robot, what is not a significant limitation in contrast to the fact that the mockup only provides one side of the simulated satellite and the robot can not perform a continuous rotation due to cable connections. Hence, tumbling satellites can not be simulated by the facility at the moment, but for testing the estimation of the target satellite pose during a servicing maneuver, this is sufficient.

There is a small model of the complete satellite available (1/30 scale), however, because the PMD camera can not measure a "scaled distance" and – more important – the light dampening including all reflections would be unrealistic, the usage of the scaled model was not pursued further.

7.1.3 Sun simulation

A 12 kW ARRI theater lamp has been used as a sun simulator and placed at different positions for the experiments (see section 7.5 on page 182), to simulate different lighting situations encountered in orbit.

The spectrum of the lamp was measured using a B&W Tek SpectraRad Xpress spectrometer (BSR112E-VIS/NIR). This particular spectrometer is specifically suited for spectral irradiance measurements. Its wavelength range is 250–1,050 nm. Figure 7.3 shows the spectrum of the lamp, compared to the sun spectrum.[1]

What can be seen is that for a comparable light intensity, high-power illumination is necessary. A comparable sun spectrum is achieved with the 12 kW lamp (ARRI-MAX) at distances less than 7 meters. There are still slight differences – and most prominently, several peaks – in the spectrum, which are related to the inner lamp gas composite. What can also be seen is the presence of an ultra-violet filter, as the spectrum is cut at about 400 nm. The 5 kW lamp is of a different architecture and does not reach comparable light levels.

Before using this plot for judging the sunlight similarity, it is also advisable to actually have a look at the sensitivity plot of the sensor, what is given in figure 7.4. Here, it can be seen that the spectral deviation of the light source from the sun above wavelengths of 1 μm can be neglected, as the camera will not capture these deviations significantly. The same applies to the ultra-violet range below wavelengths of

---

[1] For an in-depth analysis of the solar spectrum, see [140]. The data for the plot was taken from http://rredc.nrel.gov/solar/spectra/am0/ASTM2000.html.

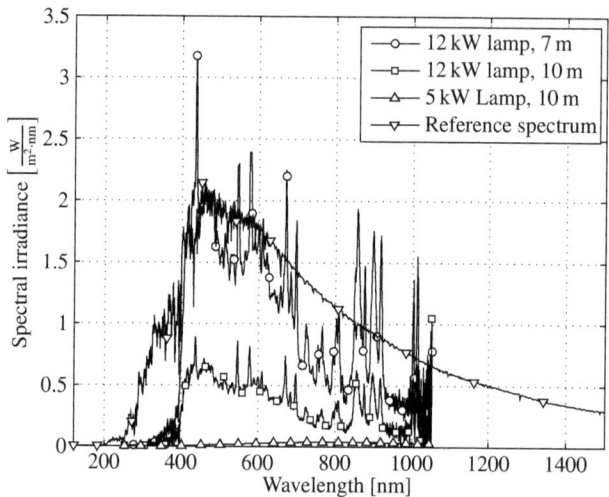

Figure 7.3. Comparison of real and simulated sun spectrum. The real sun spectrum data was taken from the ASTM E-490-00 extraterrestrial Reference Spectrum (ETR/ AM0). The spectrum of a 5 kW lamp at 10 m and a 12 kW lamp at 10 and 7 meters is also shown.

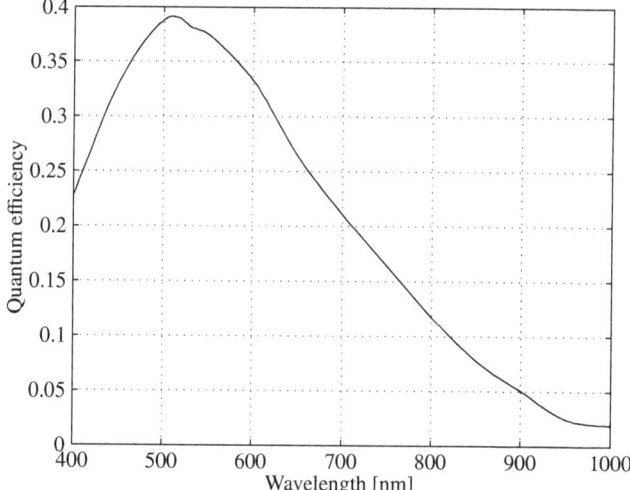

Figure 7.4. The spectral response of the CCD camera (Prosilica GC-655), as taken from the camera data sheet.

Table 7.1. Comparison of absolute irradiance levels relevant to the PMD sensor.

Location	Absolute irradiance* (W/m^2)
Earth surface†	56
AM0 (facility simulation)‡	113
AM0 (reference spectrum)	120

* Integration range: 815–940 nm.
† Measured using the spectrometer on february 22, 2012 at the DLR site in Oberpfaffenhofen.
‡ At a distance of 7.0 meters.

about 400 nm. The camera data sheet does not provide information for wavelengths below, but it can be estimated that the sensitivity will continue to gradually decrease. Therefore, the range of interest is near the peak of quantum efficiency, which is located at about 510 nm. The chosen light source provides sufficient power at this point. To conclude, using the 12 kW lamp at a distance of about 7 meters is sufficiently close to the real sun exposure when the sun is to be simulated in conjunction with this particular camera.

Unfortunately, the spectral response of the PMD camera is not available. Therefore, the simulation of the sun can not be guaranteed to be close to reality. Moreover, it must be assumed that the spectral sensitivity peaks at about 870 nm as suggested in the data sheet. The wavelength of the infrared light provided by the LEDs is in fact specified to be 850 nm, however due to the temperature-related shift occurring during operation, 870 nm is more realistic. Also, the bandwidth of the LEDs is specified to be about 100 nm. When inspecting figure 7.3, it can be seen that there are peaks in the spectrum of the lamp very close to this wavelength range, although a precise readout showed that the irradiance levels of the sun and the lamp are approximately equal at 870 nm.

According to the manufacturer of the PMD camera, the optical path consists of a RG830 filter (peak transmission at 830 nm) and an interference filter, which cuts wavelengths above 940 nm. Because there are multiple peaks within this wavelength range, the total irradiance levels have been calculated by integrating the spectral irradiance functions in the relevant interval.

There are two different conclusions that can be drawn from the results (table 7.1). First, the sun simulation of the facility is very close to the reference spectrum, which represents reality. The magnitudes of absolute irradiance are comparable. Second, there is a large difference between the absolute irradiance on the surface of earth and space. Obviously, about half of the power in this range of the spectrum is lost in the atmosphere.

The consequence of this is that when the PMD camera is advertised as being "sunlight tolerant", this will very likely only apply to scattered sunlight on ground; direct sunlight in space will be more than twice as intense. This has to be kept in mind, especially when regarding the rendezvous scenarios conducted in section 7.5 (starting on page 182).

7.1.4 System structure

The structure of the measurement environment is explained in figure 7.5. Due to limits in interfaces and also limits for run-time memory allocation, it was decided to run only the most essential parts on the real time system, that is, the robot controller and the interface. By using the SimCon block, the two simulations can be connected and the EPOS feedback is transported via a TCP connection. For details on this block, please refer to [17, 113].

For facilitating the analysis of the measurements, the image data was written to files along with timestamps and EPOS reference data. For Hardware-in-the-loop tests, this can be adapted in such a way that the estimated pose is passed back to the master Simulink model to provide inputs for a guidance system or similar.

7.1.5 Sensors

The PMD sensor was operated using the provided optics, which does not allow focus or focal ratio adjustments. The CCD sensor was operated with a more advanced optics which allows focus and focal ratio adjustments. The focus was set to infinity and the focal ratio will be provided in the descriptions of the individual experiments, as it was not left constant.

Due to the fact that the two cameras are not synchronized to the EPOS facility, latency issues can occur. The typical latencies are shown in table 7.2.

What can be seen in the table is that the standard deviation of the latencies of both sensors do not differ much. This is something one would probably not expect from

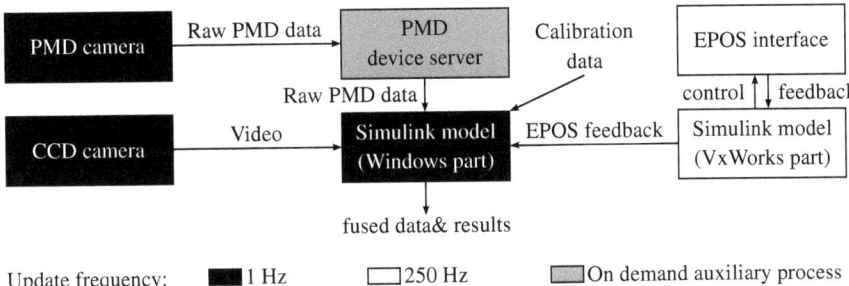

Figure 7.5. Image capturing component overview and data flow. The PMD camera only has a USB interface, which is incompatible with the real-time part (white). Therefore, a device server (gray) provides the data of the PMD camera on a TCP connection. The rest of the process is contained within the Windows part.

a setup such as this. Still, it is clear that the longer and more complex data path of the PMD camera causes the standard deviation to be larger than the one of the CCD camera. This also explains the significantly higher latency mean.

Both sensors are queried from MATLAB/ Simulink in a serial fashion, which means that the total latency of the data acquisition will be the sum of both sensors (about 240 ms).

Since the effect can not be compensated in any way and since it is very short compared to the cycle time of one second, it will be neglected in the following. Also, due to low velocities during motion capture sessions, the effect will be minimal.

7.1.6 PMD Sensor issues

There are several open questions, which must be dealt with in a dedicated fashion. The PMD sensor is mostly used and tested with a single frequency. Therefore, it is desirable to know whether the oscillator can also reliably produce other signals for different modulation frequencies. To summarize, the following issues must be dealt with first:

- **Stability of the oscillator.** The modulation frequency must be as stable and accurate as possible, because deviations accumulate and cause severe errors in the distance measurement especially for large distances.

Table 7.2. Sensor latency statistics* (measured using the performance counter API on the ACS-MMI Windows system[†]).

Property	PMD camera	CCD camera	Unit
Mean	0.150794	0.095406	s
Standard deviation	0.000882	0.000835	s
Variance	0.000001	0.000001	s^2
Minimum	0.149397	0.095266	s
Maximum	0.168779	0.109387	s

* All values include the individual exposure times, which have been set to 80 ms for the CCD camera and 2 ms for the PMD camera.
[†] Timing resolution (Windows XP limit): $\approx 2.7937 \times 10^{-7}$ s, clock source: ACPI PM timer (3.579545 MHz). Drift is unknown, but for recent systems, 50 ppm is a reasonable assumption.

- **Oscillator settling time.** Because the oscillator now must produce two different frequencies, it depends on the hardware design, whether this works flawlessly or not. Switching between two different frequencies may introduce additional problems.

- **Time between measurements.** The two measurements performed at different modulation frequencies must be made as close as possible in the time domain in order to minimize the distance error introduced for the second measurement by the delay. The delay should be constant and must be known.

In order to accommodate these issues, it was decided to take a direct approach. The Camcube provides the modulation signal for both LED cubes as an LVDS signal. By using a digital storage oscilloscope, the signal can be analyzed. Still, there is one little challenge which must be solved. Both modulation frequencies will only be visible for very short time periods; determining the frequency of the individual signals will be utmost difficult. Also, any difference between the 18 MHz and 20 MHz signal will be difficult to discover, unless the two frequencies can be observed isolated.

Fortunately, the used oscilloscope (Agilent Technologies DSO7014A) has a *pulse width* trigger mode. By forcing it to trigger for a specific pulse width, choosing two slightly different integration times at the camera, and triggering on the *length* of the active state of the LVDS lines, the two signals can be isolated. The results are shown in figure 7.6.

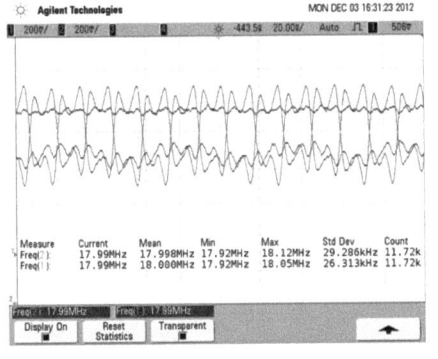 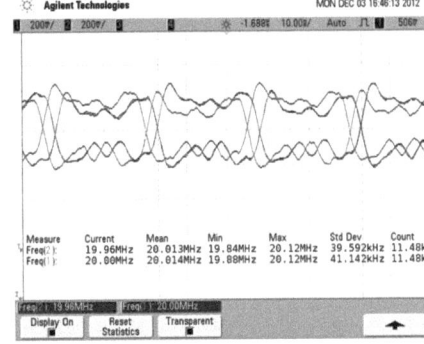

(a) LVDS signals at 18 MHz (timebase 20 μs/div)

(b) LVDS signals at 20 MHz (timebase 10 μs/div)

Figure 7.6. PMD modulation signal measurements. The complementary LVDS signals of the camera have been measured using a digital storage oscilloscope in order to determine the frequency deviations.

As can be seen, it is still very difficult of triggering at the right point. Because it now depends on a time period of a pulse (and not a rising or falling edge), the trigger point becomes a bit unstable in the time domain. Moreover, the intervals do not have clean edges for the oscilloscope to lock on to, but a very complex shape. This is why the measurement exhibits significant jitter. However, it does not pose a problem for measuring the above quantities.

The length of the intervals used for triggering correspond to the integration time. Here, an integration time of 20μs was used for the 20 MHz image and 50μs was used for the 18 MHz image. Note that the timebase is different for the two images. The modulation frequency is then measured by estimating the length of a period, what can be done automatically by the oscilloscope. Also, the corresponding frequency is determined along with statistical properties. After more than 10,000 measurements, the frequency estimate becomes sufficiently accurate.

Especially for the right image, the frequency deviation of $\Delta f \approx 0.0135$ MHz leads to a wavelength deviation of $\Delta \lambda \approx 0.005$ m. This offset will accumulate for each ambiguity interval. Therefore, the measured frequency must be used as a reference for all measurements and for the calibration as well.

What remains to be measured is the time it takes the oscillator to provide a stable signal for the LEDs and also the time between the two measurements. Both quanti-

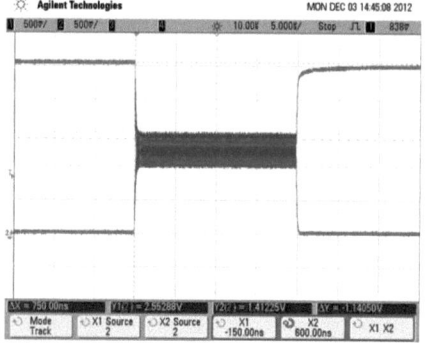
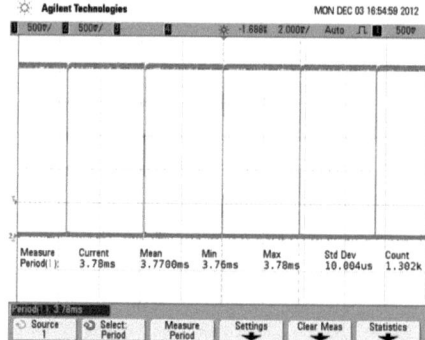

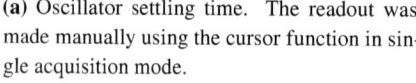

(a) Oscillator settling time. The readout was made manually using the cursor function in single acquisition mode.

(b) Time between two measurements. Measured automatically using the signal period estimation of the oscilloscope.

Figure 7.7. PMD modulation signal timing measurements. About 750 ns pass until the LVDS signal reaches a stable level once the integration begins. Between two measurements, an interval of about 3.75 ms can be found.

ties have again been measured directly at the LVDS lines of the camera. The result is given in figure 7.7.

First, the time it takes the signal to become stable was determined to be about 750 ns. For the given integration time interval of $\Delta t = 20\,\mu s$, this equals to about 3.75%. It is therefore concluded that the effect can be neglected. The beginning of the integration interval will only contribute very little to the measurement, and especially for larger integration times (which are certainly more common than 20 μs), the effect will be even less significant. The settling time does not depend on the integration time but is related to capacities on the chip, the PCB and cabling, along with inductivities.

Especially figure 7.7b shows something unexpected. When the camera is put into multi-frequency mode, the two measurements are not performed in a contiguous manner, but instead, a cool-down interval in inserted (probably to prevent an out-of-specification duty cycle on the LEDs). This way of design is highly problematic. This means that the second measurement will be performed at a significantly different point in time, what causes offsets for all non-static scenes and especially for objects moving or rotating at higher velocities, the offsets will become large.

The offset seems to be constant and not changing with the integration time, so it can also be speculated that this is perhaps not a safeguard for the LEDs but may be some time span required for internal sensor data processing (buffering, bus transport, reset/ preparation for next measurement, etc).

Nevertheless, a better design can be achieved by moving most of the processing after both measurements, so that the time between two consecutive measurements is minimized. The time between two frames (i.e., group of two images) is less relevant and restrictive than this intra-frame time span. For the given situation, the period was 3.77 ms. Subtracting the integration time, the actual processing time is at about 3.75 ms. The target object was moving towards the camera at a velocity of 5 cm/s what consequently translates into an error in distance measurement of 0.175 mm for the second integration time interval. For this given velocity, the error is negligible. However, this must be kept in mind for larger velocities.

Unfortunately, especially in cases where the integration time is set to very low values in order to become as robust as possible against motion-related imaging problems, this effect will block any progress that could be made regarding imagery of high-velocity objects. This problem must be addressed in future designs of PMD sensors.

To conclude the measurements, care must be taken with the modulation frequencies and the intra-frame spacing in time domain, when using multi-frequency imaging. Here, the effect can be ignored due to low velocities of target objects. The modulation frequency offset, however, must be respected as it has a larger impact on the accuracy. Oscillator/ signal settling time issues could not be found – their influence on the measurement is negligible.

7.2 Reference measurements

In the following, reference measurements will be made at optimal conditions, in order to better evaluate the impact of different lighting conditions. This first section is to be seen as a proof of concept, where the emphasis is put on a rather quantitative analysis of the measurement errors and their causes.

Table 7.3. Properties of the linear approach experiment (ideal conditions). The most important point for the experiment was the diffuse lab lighting, what eliminates the influence of lighting effects.

Property		Value	Unit
Target velocity		0.005	m/s
Initial distance*		22.000	m
Final distance†		1.000	m
Attitude quaternion‡	q_x	0.500	—
	q_y	-0.500	—
	q_z	-0.500	—
	q_w	0.500	—
Imaging rate		1.000	Hz
PMD camera integration time		2.000	ms
CCD camera integration time		80.000	ms
CCD sensor focal ratio		f/8	—
Lighting		Diffuse (lab)	

* The full distance of 25 meters can not be used due to the robot configuration.
† For testing the presented algorithm, this is sufficient. See the text for details.
‡ The rotation quaternion was fixed for this particular sequence.

7.2.1 Linear approach

In this measurement campaign, the target moves towards the chaser without any disturbance. Furthermore, it is attempted to achieve *optimal lighting conditions*. Optimal here means that the accuracy is maximized. This is for getting a first impression of what to expect. Table 7.3 summarizes the properties of the experiment. The main driver for this test was to see the performance of the PMD camera.

The target velocity was chosen in such a way that the experiment can be completed in reasonable time. Any approach in a real orbit between two spacecraft will – at this distance – look very similar. Higher velocities lead to an increasing risk of damaging one or both spacecraft in the docking process. Therefore, it is believed that this approximates reality good. Apart from that, by increasing the imaging rate, higher velocities are possible when required, so this is no limitation at all.

As the table states, the initial distance was chosen to be 22 meters. Although the EPOS facility can theoretically simulate a distance of 25 meters, this was chosen for simplicity. The main problem here are the two manipulators which have a configuration of their joints which needs some space for itself. For each robot, this is approximately 1 meter. The remaining difference (1 meter) was added for safety reasons. The robot configuration is flexible, but at the time this thesis was compiled, the on-board measurement system was already installed (but not yet operational), which limits this flexibility to a certain degree. Also, there are joint movement limits as well as positioning area limits in place to prevent damage to the facility and its mockups. Still, for most of the measurements, these limits have never been hit and if they have, the impact was negligible.

The final distance was chosen to be 1 meter. The presented method allows *rendezvous* of two spacecraft as it is. However, for docking, the distance becomes too small for any camera to see the entire target. In other words, target edges move out of the image what makes it impossible to detect their positions. At this point, the presented method will cease to function. This is why it is necessary to switch to a different type of sensor or algorithm at this point. For example, a feature-based approach (as done earlier in [143]) can provide estimates until the docking process is complete. Most likely, however, other sensors will be used (for example, the OLEV docking tool has sensors which can be used for the last few centimeters [66]). Multiple cameras with different field of view are also widely used [103].

The integration time was left fixed (although there are approaches of adapting it dynamically [48, 149]) mainly because changing the integration time can take a lot of time (where the camera is not responding while at it). Furthermore, the results are still reasonably accurate, as will be seen later.

The imaging rate was chosen from what is known to be very common. At this rate, image processing can take place even on embedded systems with very limited computational power, as it is the case for most satellites. Also, for a regular approach with very little velocity changes (as velocity changes are usually expensive) a navigation filter will still be very well capable of correctly performing the rendezvous [12].

The lighting for this first measurement campaign was set up very simple. The lab provides diffuse lighting through the CCFL lamps at the top of the lab. Since the lab is about 10 meters high, and the density of lamps is high, the resulting light can be assumed to be diffuse and uniformly distributed. More importantly, this lighting configuration allows testing the algorithm at peak performance, what is important for detecting lighting-related performance losses.

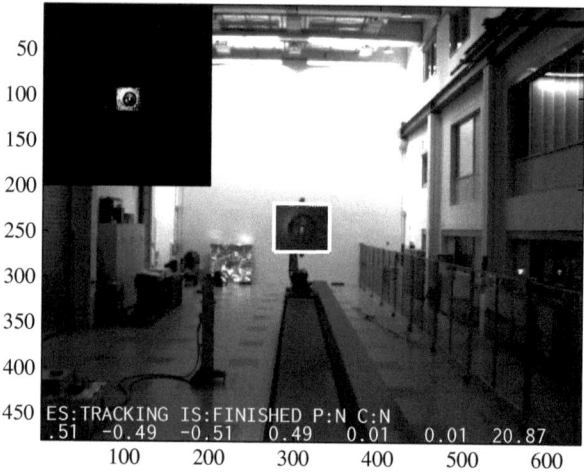

Figure 7.8. Pose estimation visualization. The straight outer edges of the model are visualized as overlay inside the CCD camera image. At the top left corner, the amplitude channel of the PMD camera is shown. Pixels which have been selected for surface estimation are mixed with 50% white. As this is being done in memory (grayscale image processing), there is no color available. The X- and Y-axes show the pixel coordinates.

A still image of the pose tracker output is given in figure 7.8 showing the CCD image with the PMD amplitude channel visualized in the top left corner during the approach sequence.

The text in the image provides the state of the algorithm: Estimator state (ES), Initializer state (IS), PMD sensor image space violation (P), CCD sensor image space violation (C). The last line provides the current pose estimate. The first four numbers are the quaternion components and the last three components form the translation vector. An image space violation is present, when at least one point of the projected target object is outside of the visible sensor area.

Slight offsets are normal, these are caused by the fact that the overlay is late by one frame. The PMD range measurement, as it is available after the PMD preprocessing (see chapter 4 starting on page 91), is shown in figure 7.9. The distance measurement of the PMD camera contains vital information about the distance to the target, as well as two of its rotational degrees of freedom. Note that the distance range was limited to 30 meters in order to improve the readability of the figure.

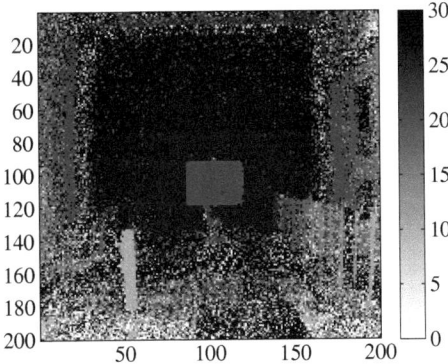

Figure 7.9. Visualization of the distance measurement. Color indicates measured distance in meters (white = 0 meters, black = 30 meters). The large amount of noise results from a very short integration time tuned for measuring the distance of the mockup. As can be seen, the mockup does not show evidence of significant noise. The X- and Y-axes show the pixel coordinates.

The result of the measurement is given in figures 7.10, which shows the position estimation error and 7.12, which shows the orientation estimation errors in form of Euler angles. The angles have been retrieved by computing the difference quaternion $q_\Delta = q_m \odot \overline{q_{ref}}$ between the reference orientation q_{ref} and the estimated (measured) orientation q_m, and then converting the difference quaternion to Euler angles using pitch – yaw – roll order, as described in section 1.6 on page 30.

While the X- and Y- directions are determined using the CCD sensor to the largest degree possible, the Z direction is determined using the PMD sensor. As can be seen in figure 7.10, especially the Z direction shows a striking error. This error can be described as having several components. First, there seems to be a periodic effect. This could be a residual from the Wiggling effect at first glance, but the residual plots of the calibration did not show any evidence of this (figure 7.11). It is possible that it is caused by unknown reflections, but despite a significant effort put into finding the cause, it still remains unclear. A few outliers have also been captured, which raise the error to about 5–6 centimeters, but they do not pose a real problem, as such incidents can be filtered out in a later stage.

The second effect is a general decrease of the error with decreasing distance (so the distance of the target is slightly over-estimated for larger distances). This is very likely a residual effect of the intensity influence on the measurement.

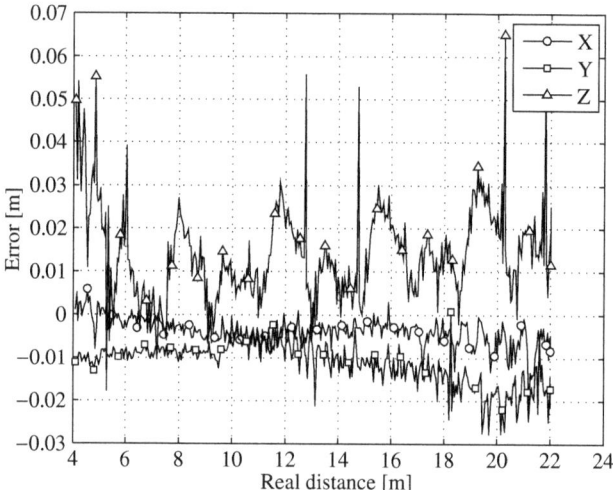

Figure 7.10. Position estimation error for the linear approach. The Z axis is target-pointing, meaning it encodes the distance to the target.

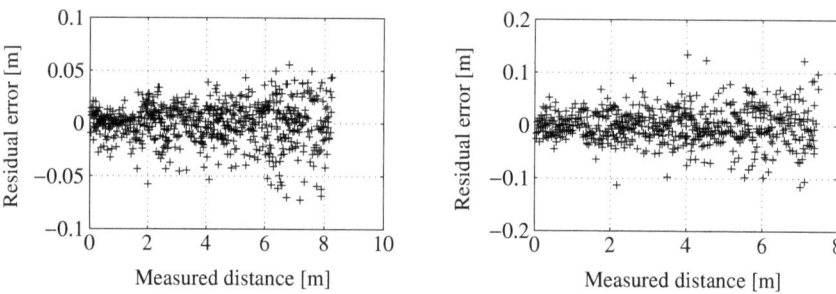

(a) Residual error after the calibration of the Wiggling effect for 18 MHz ($\lambda/2 \approx 8.3$ m).

(b) Residual error after the calibration of the Wiggling effect for 20 MHz ($\lambda/2 = 7.5$ m).

Figure 7.11. Wiggling effect residuals after compensation. In these plots, no other compensation measures are considered. Both plots indicate that the periodic error has been neutralized.

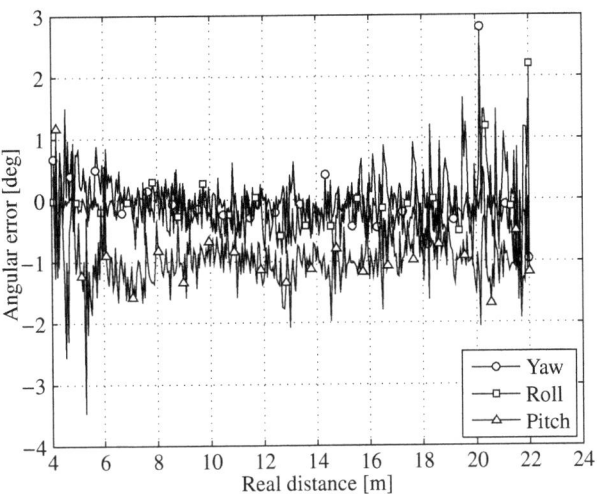

Figure 7.12. Angular estimation error for the linear approach.

Please note that details about PMD sensor calibration are out of the scope of this work. Granted, the accuracy of the estimation can still be improved by using more sophisticated PMD calibration methods. However, the range error constraint of 1%, which is commonly found in space-related navigation and distance estimation according to [130], is not violated except for very close distances below 6 m. At this distance, switching to a different pose estimation method would be advisable also for continuous tracking, as the outer edges of the target will be outside of the image space very soon. Consequently, the performance is sufficient even in its current implementation. The increase of the error for very low distances is likely to be a consequence of multi-path effects.

The angular error is shown in figure 7.12 in the form of Euler angles. What can be seen from the plot is that there is a static offset on the orientation of the target (see the one degree offset on the pitch angle), what is at first glance very likely to be a side effect of the fixed pattern noise calibration (see section 4.1.1 on page 94). As the camera was positioned against a white wall, the precision of the rotation and position is very limited. Unfortunately, there was not any suitable equipment available to achieve a higher accuracy in this regard. However, disabling FPN correction did not improve the result, so there must be a different source of error.

As the deviation remains constant over the entire distance range, it is possible that this is due to a remaining error on the extrinsic camera calibration, but this is rather unlikely because the error would have to be significantly smaller in magnitude. Most likely, it may be the case that this is a residual of a compensation measure applied by the chaser robot. The robots compensate the static effects of mass mounted near the end-effector. As the chaser was not moving, gravity would lead to a deviation in the pitch angle. The problem is now that the load parameters are configured for the target model, which was not mounted during the measurements (because otherwise the PMD camera could not have been mounted at all). Consequently, the robot was over-compensating for a mass that was not there. It is uncertain, whether this can explain the complete offset, but it is at least a part of it and it is conspicuous that the error appears on the pitch angle, which describes the up/down directional rotation and thus is sensitive to gravity. Unfortunately, it was not possible to perform a new set of measurements with the static load compensation disabled or corrected due to time constraints and facility reservations.

7.2.2 Compound motion

In this section, the target exhibits a compound motion, where all rotational axes along all translational degrees of freedom are changing with time. This test is intended to demonstrate the method for freely moving and rotating or tumbling objects. The experimental setting from section 7.2.1 is reproduced except for the distance and motion parameters. In addition to the distance-decreasing rendezvous trajectory, a disturbance was added.

The X- and Y- displacements are sine functions with an amplitude of 10 cm. The attitude quaternion is constructed from a rotation matrix, which rotates about all three angles with the same sine functions (amplitude 10°). The purpose is to demonstrate the suitability of the algorithm in more complex situations.

The results of the compound motion show that the regular pattern seen in the distance measurement of the linear approach (figure 7.10) is still partially visible (figure 7.13).

An additional problem encountered in this compound motion case is the specular reflection of the target mockup. The LED emitters of the PMD camera are clearly visible as a bright circle on the target, moving over the target as it rotates. However

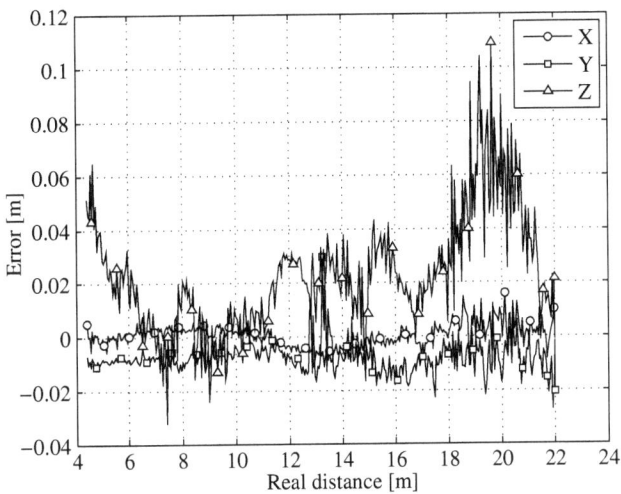

Figure 7.13. Translational error of the compound motion. The largest deviation remains in the Z direction, but the regular pattern seen in the linear approach is almost gone. Overall, the translational errors are bound by 3 cm in X and Y direction and 11 cm in Z direction.

still, despite the locally distorted distance measurements, the target attitude and position are estimated reasonably well.

When inspecting the statistical properties of the errors (table 7.4), a few things become visible as well: First of all, a problem with the Y axis becomes evident for the translational quantities. This is probably due to the static load compensation of the robots. The increase of the error in Z direction is caused by the large amplitude deviations encountered during the approach. The X axis position is estimated only slightly worse when compared to the linear approach.

The estimated orientation shows only marginally larger errors. This is due to the initial pointing error and the effect of the additional rotation. A more detailed view of the error is given in figure 7.14.

While the error shows the expected regularity, there is also a point where the orientation can not be determined reliably any more (near 20 m). This particular problem is caused by the reflection properties of the foil surface. Even small tilt motions can lead to large signal strength deviations, which manifest in large amplitude variance (figure 7.15).

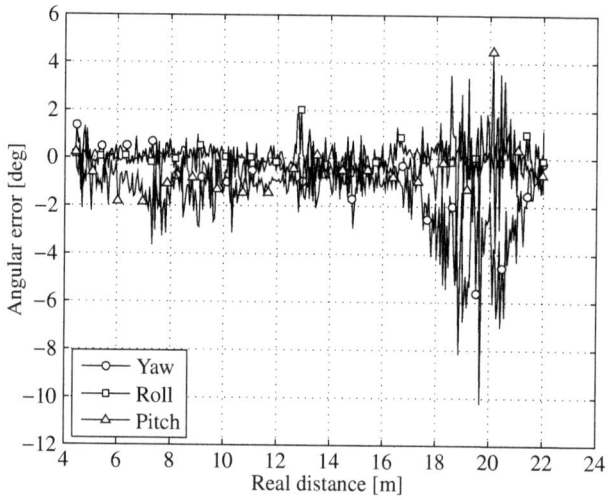

Figure 7.14. Angular errors of the compound motion. The static load compensation introduces an orientation error which manifests in deviations between the reference and the measurement.

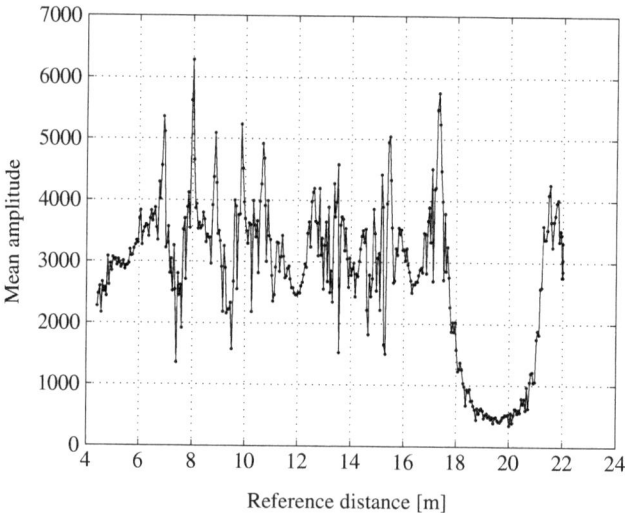

Figure 7.15. Compound motion mean amplitude (signal strength). From the set of target associated pixels, the mean amplitude is computed. When the target is tilted in such a way that the reflected signal becomes very weak, the point cloud becomes sparse and all quantities measured by the PMD camera become inaccurate.

Table 7.4. Comparison of pose estimation errors for different motions. The compound motion shows significantly larger errors due to the more complex movement. The largest impact comes from the heavy changes in lighting due to the rotation and the reflection properties of the target surface.

Property	Linear approach*	Compound motion*	Unit
Positioning error (X)	−0.004 ± 0.003	0.000 ± 0.004	m
Positioning error (Y)	−0.011 ± 0.005	−0.008 ± 0.006	m
Positioning error (Z)	0.014 ± 0.011	0.023 ± 0.021	m
Pitch angle error	−0.970 ± 0.451	−0.730 ± 1.060	deg
Yaw angle error	−0.018 ± 0.452	−1.074 ± 1.672	deg
Roll angle error	−0.133 ± 0.432	0.021 ± 0.479	deg

* All values are to be read as $\mu \pm \sigma$ (mean ± standard deviation).

This obviously happens only for distances larger than 16 meters. For smaller distances, the mean signal strength remains at an acceptable level, despite the fact that the target orientation is still constantly changing and passing through comparable situations.

In summary, this is a problem which can be ameliorated with constantly tuning the integration time of the PMD camera, however, there are still limits (the image captured by the PMD camera will always require a high dynamic range sensor).

Regulating the integration time can be kind of dangerous in situations of rotating targets, as the amplitude values change very quickly (see the rise and fall of the amplitude near 18 and 19 m in figure 7.15). In the worst case, this may even lead to tracker divergence and when the initializer is used to reinitialize the tracker, it may not be able to find the target due to a large number of saturated pixels. Therefore, it is not sufficient to just adapt the integration time to the mean amplitude of the target pixels, as it was proposed before [48, 149]. Instead, one must also hold and propagate a *safe* value, which may be used in this situation in order to prevent tracker divergence or reinitialization problems. This safe value would have to be chosen in such a way that target acquisition is always possible (maybe with limited accuracy, but this would be still better than not finding the target at all). Apart from that, there is a significant time delay between the integration time change request and the actual resulting image. An integration time controller would have to take this into account.

Table 7.5. Static pose estimation error statistics. Four different distances have been used to identify possible laws regarding target distance and estimation quality. The integration time (2 ms) and environmental lighting (diffuse lab light, as described in section 7.2.1 on page 162) were fixed for these measurements.

Pose component	5 m*	10 m*	15 m*	20 m*
Position (x)	0.003 ± 0.001	−0.001 ± 0.000	−0.002 ± 0.001	−0.001 ± 0.004
Position (y)	−0.009 ± 0.001	−0.008 ± 0.001	−0.008 ± 0.002	−0.009 ± 0.006
Position (z)	0.025 ± 0.006	0.006 ± 0.002	−0.006 ± 0.004	−0.013 ± 0.017
Pitch angle	−1.026 ± 0.505	−1.099 ± 0.142	−1.216 ± 0.301	−1.157 ± 0.467
Yaw angle	0.250 ± 0.372	−0.168 ± 0.144	−0.173 ± 0.369	−0.256 ± 1.479
Roll angle	−0.056 ± 0.092	−0.000 ± 0.081	−0.049 ± 0.137	−0.220 ± 0.422

* All values are to be read as $\mu \pm \sigma$ (mean ± standard deviation). Sample size: 90 (for eliminating startup effects, the last 90 frames have been used). Position values are in meters, angles are in degrees.

7.3 Static target performance evaluation

For this test, the target object has been positioned at different distances (because the distance is the most important parameter that influences the quality of the measurement) and for each point, about 100 frames have been captured. The resulting mean values and standard deviations are shown in table 7.5.

The static target analysis is intended to provide an insight into the influence of the distance, while preventing motion-related effects from distorting the statistics at the same time.

As can be seen from the table, there is an increase of position uncertainty for the X- and Y- directions due to the varying size of the target. The more far the target moves away from the chaser, the more influence is gained for a single pixel. This can also be seen in the standard deviation of the roll angle, as it is related to the same effect.

The distance estimation is rather independent of the distance, as expected. The only influence here is the signal dampening, which depends on the distance. However, its influence on the standard deviation is still very limited and from the series given in the table, no concrete law can be derived.

For these measurements, the initializer was supplied with a surface rotation hint, in order to prevent random object rotation configuration ambiguities from complicating the comparisons. Note that this does not happen during a continuous sequence, as local attitude information is used in order to resolve these ambiguities.

The angular offset on the pitch angle is visible again in form of a static offset of about $-1°$, as already explained in section 7.2.1 on page 162. Apart from this, it can be generally concluded that the roll angle can be determined with a slightly higher accuracy than the two other angles. Especially for low distances, near-range effects related to the PMD sensor amplify this effect, as there are increasing problems with determining the angles from the point cloud. On the other hand, at large distances, the roll angle estimation becomes less accurate and there may be a point where it becomes less accurate than the pitch and the yaw angle. This is related to the differences in lateral sensor resolution and depth resolution.

7.4 Start-up process evaluation

In this section, the start-up of the presented algorithm is evaluated. To that end, a few different scenarios are used in order to compare the initializer pose estimate with the reference pose. Also, it is investigated whether the tracker did correctly converge on the initial pose estimate.

7.4.1 Initializer performance

Given the architecture of the initializer (see chapter 5 starting on page 115), the quality of the initial estimate depends solely on the PMD sensor. In the following, different scenarios are evaluated in order to provide a thorough analysis of the performance of the initializer.

Static case

First, the four static scenes already introduced in section 7.3 will be used. The results are shown in table 7.6.

What applies in general is that the standard deviations are larger compared to those of the tracker (table 7.5). This is due to the lack of information from the CCD sensor. Also, the standard deviations increase with the distance as expected, since less pixels are contributing to the estimation when the target appears smaller.

Table 7.6. Static target acquisition error statistics for targets at four different distances (same image material as used in section 7.3). A surface rotation hint (determining $\mathbf{q}_{rot}$ as described in section 5.2.4 on page 122) was provided to the initializer here, in order to facilitate the computation of the mean and standard deviation values.

Pose component	5 m*	10 m*	15 m*	20 m*
Position (x)	0.165 ± 0.017	−0.104 ± 0.004	−0.183 ± 0.005	−0.245 ± 0.005
Position (y)	−0.041 ± 0.007	−0.015 ± 0.003	−0.010 ± 0.004	−0.002 ± 0.004
Position (z)	0.061 ± 0.012	0.003 ± 0.007	−0.007 ± 0.005	−0.020 ± 0.011
Pitch angle	1.204 ± 0.842	−1.073 ± 0.055	−1.251 ± 0.056	−1.093 ± 0.082
Yaw angle	0.061 ± 0.451	−0.308 ± 0.057	−0.237 ± 0.069	−0.225 ± 0.089
Roll angle	−24.165 ± 2.301	2.486 ± 0.587	3.873 ± 0.904	1.821 ± 0.868

* All values are to be read as $\mu \pm \sigma$ (mean ± standard deviation). Estimates marked as invalid have been filtered in order to retrieve correct statistical properties (frames for which the initializer was unable to provide an estimate). Position values are in meters, angles are in degrees.

The in particular bad estimation of the attitude quaternion at 5 meters is again likely to be related to near-range effects, what demonstrates in a non-intentional way the robustness of the tracker (figure 7.21, page 179). It is caused by a significant deviation of the target object shape as it is represented in the resulting target affinity matrix produced by the initializer (figure 7.16).

Dynamic case

For testing the initializer during motion, a reference trajectory has been used, where the distance between the chaser and the target is continuously reduced and at the

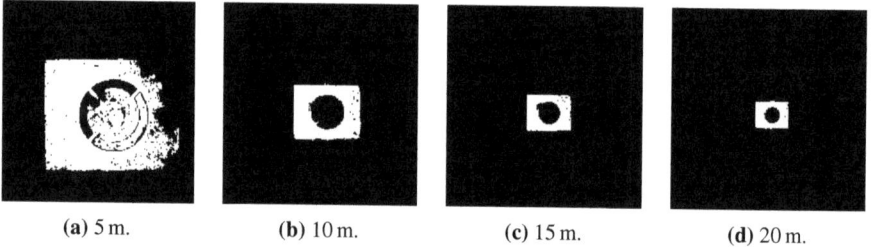

(a) 5 m.   (b) 10 m.   (c) 15 m.   (d) 20 m.

Figure 7.16. Target affinity matrices for different distances. White pixels are included in the point cloud, black pixels are considered not to be a part of the target object. The target is deformed at close ranges, leading to increasing errors in the target acquisition result.

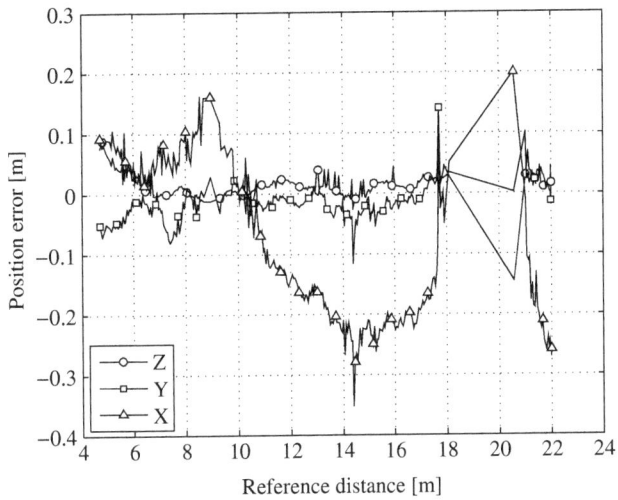

Figure 7.17. Target acquisition analysis (position).

same time, the target changed its orientation. A sine function with 10° amplitude has been added to all rotational axes and further, sine functions with an amplitude of 10 cm have been added to the two remaining translational axes. The tracker has been disabled for this test case, so that the resulting pose estimate reflects the result of the initializer for every single frame of the image sequence.

When inspecting the position estimation error of the initializer (figure 7.17), it becomes obvious that orientation changes (and consequently, severe changes in amplitude due to non-lambertian reflectance of the thermal insulation) have a large impact on the accuracy of the centroid position determination, since the centroid position is primarily used for determining the target position (see section 6.2.1 on page 120). Although the absolute errors appear to be large, when putting this into context with the distance, the errors do not exceed 5 % of the distance.

Furthermore, as will be seen later, the tracker will still be able to lock onto the target, despite these offsets. For future improvement, instead of estimating the centroid of the point cloud, fitting the surface into the point cloud will probably provide a more accurate result.

When inspecting the orientation plot, there is evidence of significant error (figure 7.18). This is the consequence of the PCA being relatively sensitive to local

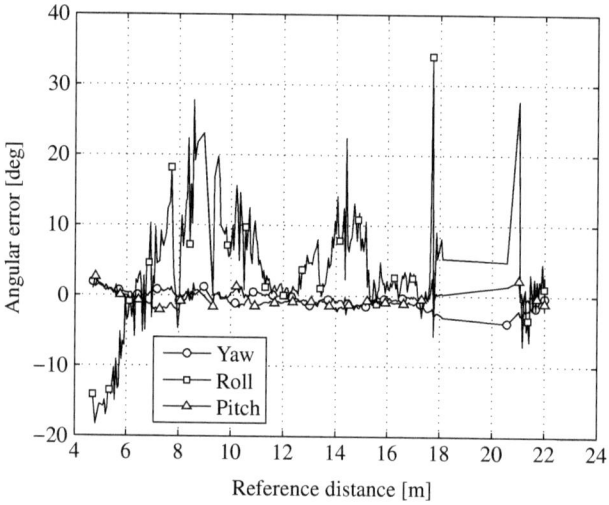

Figure 7.18. Target acquisition analysis (orientation).

PMD point cloud sparseness. The trajectory used is particularly difficult for the initializer, because it contains parts where the target does not reflect much of the signal (figure 7.19a).

While the case of the low amplitude does not pose a large problem in space, it does for the lab tests, because local reflections at the facility safety fence can confuse the tracker (false target). When the bootstrap point is placed at such a location, none of the surfaces of the model will match. Fortunately, this can not happen in space.

A more relevant problem can occur at low distances, when the target has highly articulated structures on the visible surface which produce closed areas of high amplitude and similar distance (figure 7.19b). In these cases, the point cloud can not be extended past the structure boundary, what again leads to the problem that no model surface will match.

There are also parts in the plots, where the initializer was unable to provide an estimate at all. This happened once during a low-signal phase (distance range 18–21 m) and again (but much less severe), when the target was already very close to the chaser (distance range 5–7 m).

The reason for this is that the relative matching of target pixels dropped below the threshold (figure 7.20), which has been set to a value of 0.3 here, what has

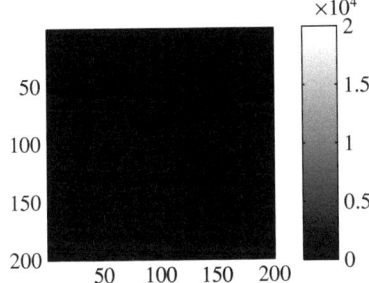 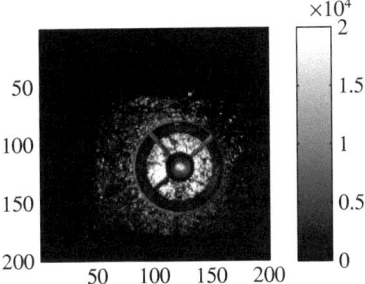

(a) Problem #1: A very low amplitude occurring due to non-lambertian reflectance properties of the target under rotation.

(b) Problem #2: Closed point set. When the bootstrap point is located within a closed structure, the point cloud can not be expanded to the full target.

Figure 7.19. Initializer image amplitude scenarios. Both scenarios pose a problem to the initializer. Either very low amplitude or high amplitude inside a closed structure can prevent the initializer from expanding the point cloud. The X- and Y-axes show the pixel coordinates.

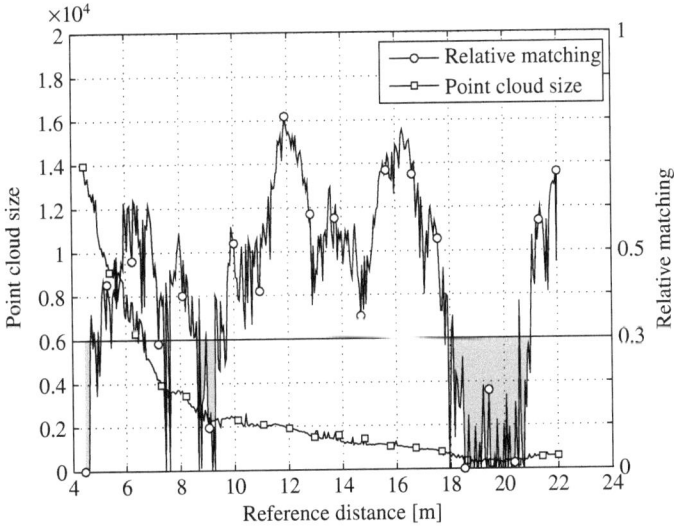

Figure 7.20. Initializer state variable progression. The curve marked with the squares shows the number of points in the point cloud determined by the initializer. The other curve shows the relative matching (1.0 means all points of the point cloud have been matched). When the matching threshold of 0.3 is not reached, the initializer will not return a pose estimate and switch to an error state. For better readability, this is marked gray.

been determined experimentally. The figure also shows the number of points in the target object point cloud. As can be expected, it increases, as the distance becomes smaller. Especially for very low distances, the effect shown in figure 7.16 limits the expansion of the point cloud and thus, the pixel-target matching becomes unstable.

As can be seen, the relative matching correlates to the point cloud size. Near a distance of 20 meters, the size of the point cloud decreases slightly, as the bootstrap point becomes unstable for several frames. Near 9 meters a significant amplitude spot causes the point cloud expansion to exclude several pixels. As a consequence, the principal components become more inaccurate and the relative matching decreases rapidly. At 6 meters and less, the point cloud can more and more not be reliably expanded to the full target size any more due to the effect shown in figure 7.16, and despite the still increasing number of pixels in the point cloud, the centroid estimation becomes inaccurate, what causes the decreasing number of matched pixels.

Overall, the approach works very well for a large target range, but still needs improvement for very low distances. The problems at larger ranges are a consequence of side effects only present in the lab experiment.

The accuracy of the initializer is only partially required for the application, as stated in the requirements defined at the beginning of the chapter. The question is whether the estimate is accurate enough for the tracker to converge on such an estimate. This will be investigated in the following subsection.

7.4.2 Tracker convergence

With the given architecture, it is assumed that the target object can be tracked, as the position changes and the angular changes from one frame to the next are sufficiently small. This translates to: the tracker will converge on an initial pose estimate, when its deviation from the real pose is sufficiently small. However, as stated earlier, the target acquisition process is particularly difficult at close ranges due to various effects, such as maximization of optical distortion effects, possible PMD multi-path effects, and so on. In this subsection, it will be investigated whether the performance of the initializer meets the requirements of the tracker.

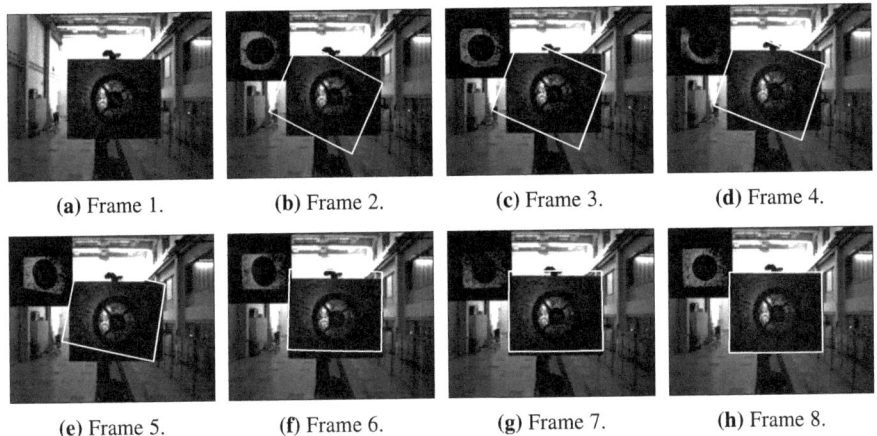

Figure 7.21. Tracker convergence after bad initialization. As the image sequence shows, the tracker is capable of correcting large errors of the initial pose estimate after a few frames.

Near-limit test

The first test was motivated from the image sequence obtained during the initialization of the static 5 meter sequence, which is shown in figure 7.21.

This sequence is particularly interesting, because it takes the tracker several frames until it has finally converged on the real position. As can be seen, at the beginning, there is a large angular error remaining from the initializer. However, the tracker was able to successfully lock onto the target. Still, in situations where lighting is critical, such a large displacement can become an obstacle for the tracker. An exemplary scene will be given later in section 7.5.2 on page 185: Under these circumstances, the CCD sensor can not provide much information, as there is not enough light available which would allow edge tracking. The initializer, however, would continuously provide valid estimates, as the PMD sensor can operate without dependence on environmental light. Nevertheless, the image sequence just shown is a first indication that the tracker is very robust. This motivates to investigate it further, specifically for a more advanced trajectory.

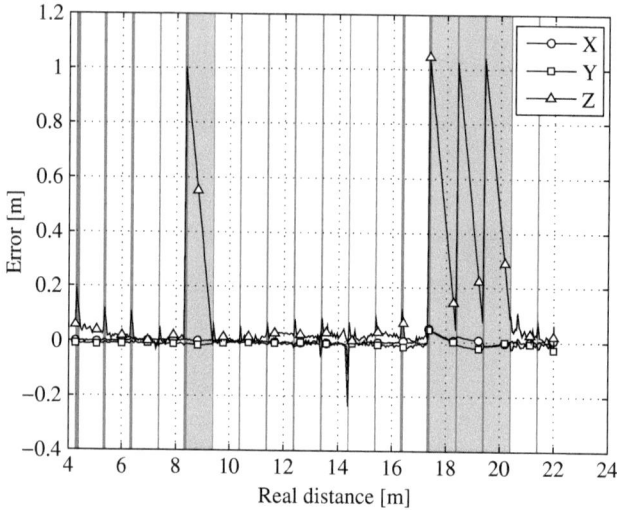

Figure 7.22. Tracker convergence during the approach sequence (position error). The background coloring indicates the estimator states (light gray: failure, dark gray: target acquisition active, white: tracking).

Complex trajectory

In the following, the test trajectory already introduced in section 7.2.2 (page 168) will be used in which the distance between the target and the chaser is continuously decreasing and the target is changing its orientation using the sine function on all rotational axes with an amplitude of $10°$. Every 20 frames, the tracker is intentionally reset, in order to restart the target acquisition and lock-on process.

When inspecting figures 7.22 and 7.23, the regular pattern where the tracker has been purposedly reset can be seen as dark gray lines. Whenever the estimator switches to the target acquisition state, the target was artificially "lost" shortly before.

Both plots show blocks in light gray, where the initializer was unable to provide a reasonable pose estimate and switched to its error state. The two cases are identical to those already explained in section 7.4.1 on page 173. Apart from these special cases, the tracker converges correctly on all pose estimates provided by the initializer.

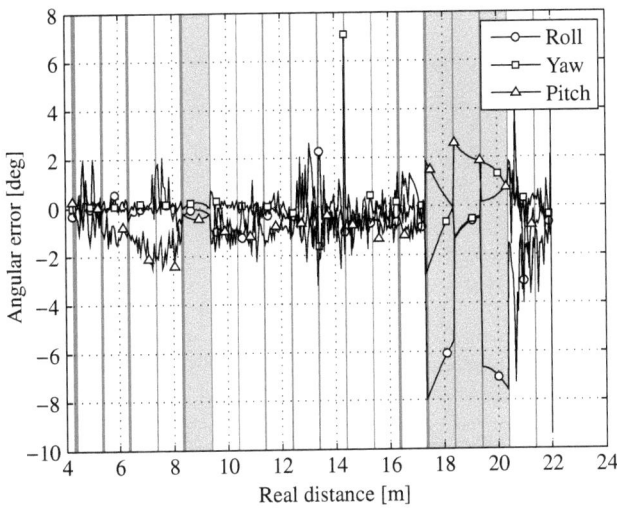

Figure 7.23. Tracker convergence during the approach sequence (orientation error). The background coloring indicates the estimator states (light gray: failure, dark gray: target acquisition active, white: tracking).

In the position error plot, small peaks can be seen for the Z coordinate, which are about 5 cm in size. Sometimes, the deviation is larger, but the tracker still converges. This can be caused by wrong inclusions or exclusions of pixels into the target point cloud used for the estimation of the pose. What can also be seen is that the position error is slightly larger also for the regions where the tracker is running. This is a consequence of the much higher dynamic range of the active PMD signal, which is now occurring due to continuous rotation of the target during the approach.

When inspecting the plots near a distance of 14.5 meters, a large orientation deviation is encountered, which is corrected within one frame. The remaining parts of the plots are consistent with the expectations (cf. figures 7.13 on page 169, and 7.14 on page 170).

Overall it can be concluded at this point that despite the difficulty the initializer might have of estimating a valid initial pose, when it does, the tracker converges to the correct pose estimate within a few frames. Obviously, the initial pose estimate must be very close to a false local optimum in pose space in order for the tracker to hold on to it by mistake.

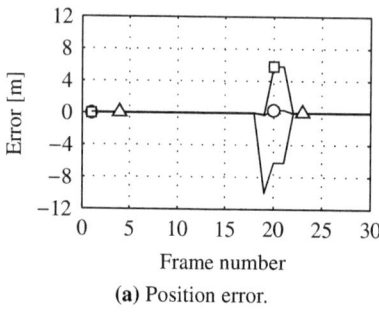 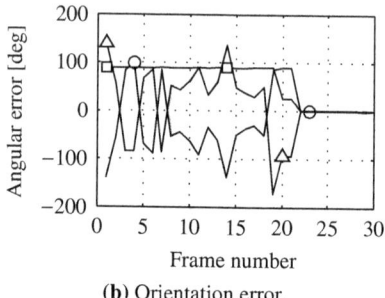

(a) Position error. **(b)** Orientation error.

Figure 7.24. Erroneous target acquisition recovery. When the target object is mistakenly acquired with a pose which is partially stable, but still significantly wrong, the tracker diverges after a brief period of time, and a new acquisition is initiated, what will eventually recover the true pose estimate. The markers correspond to the individual degrees of freedom (circle: X and pitch, square: Y and yaw, triangle: Z and roll).

Erroneous target acquisition recovery

It is possible to create such situations intentionally, for example a rotation of 90 degrees about the roll axis will cause the tracker to follow the wrong pose. However, these pose estimates tend to be unstable and it is very likely that the tracker diverges soon (figure 7.24), so that a new initial pose can be acquired. Furthermore, if there is any cross-checking available from the application side, the estimator can be instructed to re-acquire the target at any time.

This completes the synthetic performance tests. Now, an artificial sun light (as introduced in section 7.1.3 on page 153) is added in order to evaluate the impact of strong environmental light.

7.5 Rendezvous scenarios

In the following, three different rendezvous scenarios will be investigated. The difference between the experiments will be the position of the simulated sun, in order to see the impact of different lighting situations.

The position of the sun is dependent on the mission and the resulting requirements for the instruments. There are three extreme cases of sun illumination, which will

Figure 7.25. Rear sun illumination, as seen from the CCD camera. The X- and Y-axes show the pixel coordinates.

be investigated here as placeholders for a complete evaluation. Consequently, this section should be seen as a rather qualitative approach.

7.5.1 Backside sun illumination

When the sun is located behind the servicing spacecraft, but still in such a way that the servicing spacecraft does not cast a shadow on the target, the most convenient situation is found. The target is illuminated by the sun, a lot of features are visible and the background remains dark. It can be speculated already, that this will most likely produce the best results. A snapshot of the sequence is shown in figure 7.25.

Here, the target is well visible for the CCD sensor due to the scattering of light on the surface of the target. With the correct focal ratio and exposure time, it is possible to retrieve a good image. However, it would show that the PMD sensor has – despite the suppression of background light – trouble providing a clean image. In fact, the image does not allow pose estimation at all. Apparently, suppression of background light (as it is available with the PMD camera) means scattered light on earth's surface. Sunlight in space and especially direct sunlight (with corresponding intensity) do, unfortunately, not seem to be targeted by this technology.

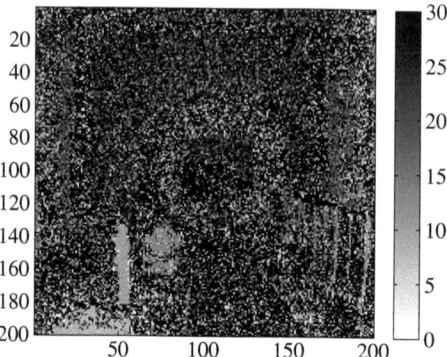

Figure 7.26. Rear sun illumination, as seen from the PMD camera (distance image, distance in meters shown in the scale on the right). Strong sunlight saturates the PMD pixels, making distance measurements impossible. The X- and Y-axes show the pixel coordinates.

The distance image of the PMD camera is shown in figure 7.26. Obviously, the PMD sensor has severe problems when dealing with large dynamic range scenes such as these. As can be seen from the image, the noise level is very large. The target is still visible in parts, but the distance measurements are completely erratic. So this image can not be used for tracking the object.

Multiple attempts of retrieving a usable image have failed – either the pixels of the PMD camera were saturated when the integration time was set to a high value, or the remaining signal was too faint when the integration time was set to a low value. Unfortunately, the integration time is the only parameter which can be modified in order to adapt to the situation. But since decreasing the integration time also decreases the sensitivity of the camera for the modulated light and hence, has a bad impact on the signal, there was no way of solving this issue. For a possible hardware solution, see section 8.4.4 on page 206.

As will be seen later, this situation has the unique property of being advantageous for one sensor (the CCD camera) but problematic for the other sensor (the PMD camera). Because the fused approach relies on both sensors, it is sensitive to this situation. While the suggested hardware change is one way of solving it, another way may be using a purely CCD-based pose estimation method, such as the one presented earlier in [142].

This shows that having multiple methods available and the ability to switch between them would also be an invaluable feature which provides robustness and flexibility to an optical navigation solution. Unfortunately, one would have to know when to use which method. In this case, it is probably easier to make changes to the hardware.

7.5.2 Side sun illumination

When the sun is located to the left or to the right of the satellite constellation, light does not scatter on the target surface, what causes it to become dark. This is very difficult for any vision-based pose estimation approach without active illumination. In this case, the LED emitters of the PMD camera also provide some light for the CCD camera, but due to modulation, the brightness of the target changes constantly as a consequence of the interference between the shutter of the CCD camera and the modulation of the infrared light.

For testing this scenario, the distance of the target was fixed at 10 m. The target was not moved, but the light source was slowly rotated. Its position was fixed at about 4 m left of the target (as seen from the chaser camera) and then it was slowly rotated in order to simulate a changing position in orbit.

It would show that the PMD camera is very advantageous in this situation. Since the light is now not visible (at least, for the most part), the pixels of the PMD camera are not saturated and hence, provide reliable distance measurements (figure 7.27). The initializer is thus capable of continuously providing rough estimates.

However, the situation is much more challenging for the tracker. At first, the target can be tracked, but as the light changes slightly, the tracker pose estimate becomes inaccurate until it finally diverges. After re-initialization, the estimate does not seem to be as stable as it was before (figure 7.28).

The problems encountered by the tracker are the result of a very low exposure time of the CCD camera (10 µs) and also a small focal ratio (f/16) required due to the high intensity of the simulated sunlight. Despite the degraded performance and the fact that the target is almost invisible to the CCD camera, the tracker is still able to at least track the target for brief periods of time. This demonstrates very well the superiority of the texture segmentation approach along with the scanline quantization (section 6.3.2, page 139), although it does not work without problems.

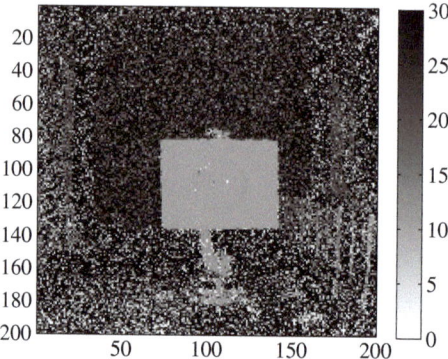

Figure 7.27. Side sun illumination. Reliable distance measurements are possible, as the overall light intensity is sufficiently low. Distances are in meters, the scale is shown on the right. The X- and Y-axes show the pixel coordinates.

Interestingly also, this situation is contrary to the case with the rear illumination. Here, the PMD sensor can provide good measurements, but the CCD sensor runs into problems. There are two possible solutions. Either use a purely PMD-based approach, such as [143], or regulate the exposure time of the CCD camera.

Either way, this shows once more that using two sensors instead of one does *not* necessarily improve the robustness of the method, but rather increases the accuracy at the price of robustness.

7.5.3 Frontal sun illumination

When the sun is located at the front of the servicing spacecraft (and behind the target), sunlight directly hits the sensors and causes pixels hit by the sunlight to be overexposed. Saturated pixels do not provide information any more. Also, the target is completely dark, when this situation is encountered. This is, among the side sun illumination, probably the worst condition which can ever be encountered.

In the experiment, the position of the target was fixed, but the orientation of the target was dynamic. Using three rotation matrices, an attitude quaternion was composed, where each one of the Euler angles corresponding to the rotation matrices was retrieved from a sine function with an amplitude of 10° (in fact, this was the

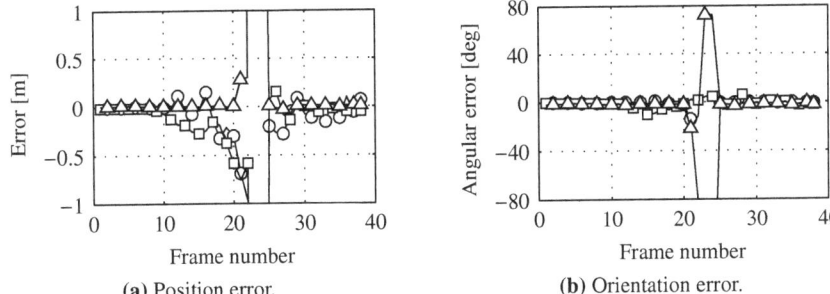

(a) Position error.   (b) Orientation error.

Figure 7.28. Tracking performance for constellations where the sun is located at the side. Overall, the accuracy degrades and even tracking failure may occur (frames 21–24), nevertheless, pose estimation remains possible. The markers correspond to the individual degrees of freedom (circle: X and roll, square: Y and yaw, triangle: Z and pitch).

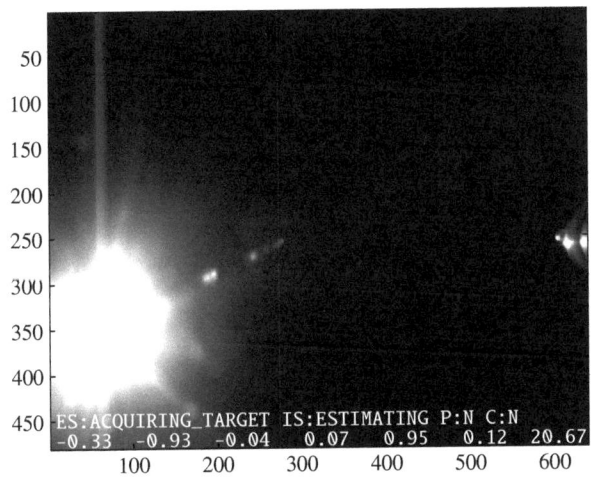

Figure 7.29. Frontal sun illumination (CCD sensor). The contrast is utmost strong; the target object is almost invisible. Under normal circumstances, the target object can not be seen, unless the chaser spacecraft reflects some of the sunlight in the direction of the target. The X- and Y-axes show the pixel coordinates.

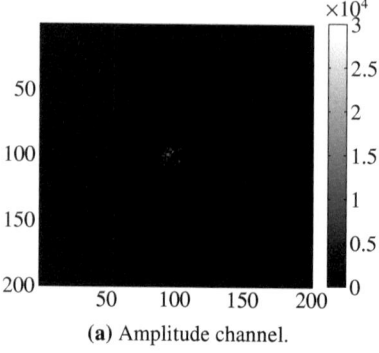 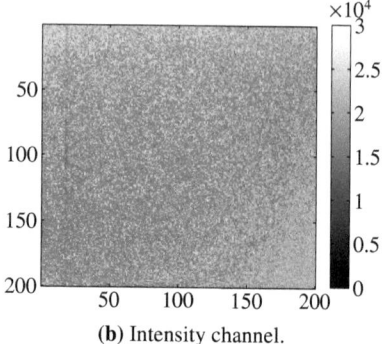

(a) Amplitude channel.   **(b)** Intensity channel.

Figure 7.30. Frontal sun illumination (PMD sensor). The strong light influences the measurement of the PMD sensor and saturates the pixels despite active SBI (darker pixels in the right figure). Saturated pixels have an amplitude close to zero. In the left figure, the target is only visible partially due to numerous saturated pixels. The X- and Y-axes show the pixel coordinates.

same motion as already used in section 7.2.2 on page 168 but without any position changes).

A typical image of the scene as captured by the CCD sensor is provided in figure 7.29. Because the target is not illuminated specifically for the CCD sensor, it appears very dark due to the small focal ratio. The fact that it appears at all is related to the side effects of various reflections of the light and in general – what the image does not show – the entire laboratory was as bright as a day in sunlight.

Attempts to start up the tracker on this image failed repeatedly, again due to the PMD camera and the effects of the strong light on the measurement. The amplitude and intensity channels of the PMD camera are given in figure 7.30.

Apparently, this light intensity is too high for the PMD sensor. When inspecting the plot of the intensity channel, it becomes evident that SBI is active: Pixels near the border show a large intensity value (light gray), while a lot of pixels in the rest of the image remain darker, indicating lower intensity values. These lower values are caused by the SBI circuitry.

This is, by itself, not a problem – however, when inspecting the amplitude image, the target object can be barely seen, as a lot of the pixels have dropped almost to zero. This means that despite the correction current of the SBI circuitry, the potential wells of the respective PMD pixels have reached saturation. As a consequence, no distance measurement is possible.

The resulting point cloud represents the target object only sparsely, which is why the initializer can not reliably find the dominant surface and thus, fails. This particular problem can only be solved by hardware means, as the CCD sensor also can not provide data leading to reliable measurements. See section 8.4.4 on page 206 for details.

7.6 Summary

In this chapter, the test environment was described. A mockup of a geostationary satellite was used in order to test the developed relative pose estimation algorithm under utmost realistic conditions. Typical rendezvous approach trajectories have been tested under different environmental lighting, and the approach motion was also modified to include various disturbances.

The pose tracker and the initialization method have been successfully tested on these trajectories, what proves that the envisaged algorithms can actually provide the measurements required for a scenario like Rendezvous and Docking.

However, the impact of sunlight is significant. In contrast to previous publications (for example, [111, 138, 143]), the simulated sunlight was significantly closer to real sunlight in orbit with regard to the spectrum and the total irradiance. One scenario has been identified where only one of the two sensors can provide valid measurements. When the sunlight originates from the back of the servicer, the CCD camera can accurately see the target object, but the PMD camera can not be used, as the light intensity is too strong (see the following chapter for a solution). When the sun originates from the back of the target, a situation is encountered, where only the PMD camera can provide measurements after hardware modifications. In the case where the sun is located at the side of the constellation, a scenario is found which is close and comparable to an eclipse phase. Here, tracking was possible – although less accurate than when operating under optimal lab conditions, as could be expected.

Overall, two future extensions seem to be necessary: a hardware modification of the PMD camera, providing more resistance to environmental light, and a software switch that allows to use only one sensor for estimating the pose, when one of the two sensors produces readings which can not be used.

In the next chapter, the results will be discussed in more detail and put into context with comparable existing approaches and different sensors.

8 Summary & Discussion

As it was seen in the last chapter, using two different sensors to retrieve a measurement can improve the accuracy, but it does not necessarily improve the robustness as well. Now, it remains to be investigated how the proposed method fits into the field of other pose estimation techniques, which may even rely on totally different sensors. In practice, probably a combination of different methods is required in order to be capable of adapting to a large range of environmental conditions.

Also, there are still concerns about the space-readiness of the PMD sensor. As the near future will be very likely to bring first experiments with PMD sensors in space, some initial thoughts about the impact of the space environment on the sensor must be made as well.

8.1 Comparison of pose estimation methods

In this section, several different pose estimation approaches will be compared. The intention is to get an impression of the current state of pose estimation methods regarding PMD sensors, CCD sensors, laser scanners and hybrid approaches based on various implementations of sensor data fusion.

8.1.1 PMD sensors

What is specifically interesting is comparing the presented approach to methods, which rely on a single sensor (either PMD or CCD) only. Using two sensors is more expensive in all relevant dimensions, therefore, significant advantages must result as a consequence. Unfortunately, very little work has been done on pose estimation with a single PMD camera so far. Much more common are approaches to environmental mapping and camera pose estimation. Nevertheless, a comparison is made where it is applicable, but often, only qualitative statements can be made due to the large differences in measurements, models and lighting conditions.

In [112], a PMD camera was used for pose estimation with an edge tracking approach. The distance information has been used to measure the orientation and also the distance. Unfortunately, the PMD camera used was not calibrated, leading to large errors. As several plots in the original work show, the 1 % rule is violated repeatedly (see for example figures 4.5–4.29). However, this is to a large degree the consequence of not calibrating the camera; the method itself would probably benefit a lot from filtered sensor data.

The camera is calibrated in [70], but there is no statement made about the accuracy. However, from stationary plots, it can be inferred that the accuracy in the distance is limited: in figure 6.21, there seems to be an error of about 20 centimeters, what is way larger that what can be achieved with the method presented in this thesis.

The work at hand also outperforms an earlier approach of using a feature-based pose estimation method [143]. It was inspired by common SLAM algorithms which usually only operate in a two-dimensional image space. Unfortunately, there was no calibrated reference measurement system available, what makes it very difficult to obtain accuracy statements. Also, noise and outliers had a large impact on the estimated pose – especially for determining the orientation.

A simulation of pose estimation with a PMD camera using point cloud matching was carried out in [56]. It should be noted that the results are not from a real setup, but from a simulation. Still, the total accuracy is comparable to the work at hand, what is good, since typical PMD sensor non-linearities and error sources are not addressed at all, meaning the calibration performed within the work at hand is sufficient.

To summarize, it can be safely said that using a CCD camera in addition to a PMD sensor is very likely to increase the accuracy. Overall, the achieved accuracy is well within reasonable borders and looks very promising, as long as PMD sensors alone still struggle with low lateral resolution and numerous sources of error.

8.1.2 CCD sensors

CCD sensors have already been used for a long time. A lot of different pose estimation methods have emerged, although it seems that single CCD sensors are rarely used. Stereo vision is much more common. In the following, however, only pure monocular vision approaches (which are designed to work with uncooperative target objects) are considered.

When using features on the target for estimating the pose of a spacecraft, a closed-form solution is possible. A positioning accuracy of about 1 mm can be achieved along with an angular accuracy of 0.3° [104]. A good reference that shows what is possible under ideal conditions when using a single CCD camera is also given in [154], as the results are based on a simulation only. Errors well below 1 mm for all translational degrees of freedom and errors in the order of 1° for the rotational degrees of freedom can be achieved.

A very interesting approach is made in [9] (however, for limited distance ranges only) that uses several different monocular cues (such as lighting, defocus) as hints for obtaining 3D positions of points and then, by fitting a model, the complete pose of a known object. An OpenGL rendering of the object is used as a comparison with the image and a suitable distance metric is minimized in order to iteratively refine the pose estimate.

Finally, an earlier approach [141] nicely demonstrated the most important problem with monocular pose estimation approaches. Overall, the CCD sensor based pose estimation is highly accurate for some degrees of freedom (see the reference for details) and certain distances, however, the performance degrades rapidly for increasing distances. Also, there is a significant dependence on which pose properties are estimated. The distance to the target object, for example, is very difficult to determine. The same applies to for the out-of-image-plane rotation angles.

Summarizing, CCD sensors are the most simple way of implementing pose estimation hardware, but have severe drawbacks in certain situations. This is the main reason for the implementation of the pose estimation using sensor data fusion. For further reading, a good overview of CCD sensor based pose estimation is provided in [82].

8.1.3 Laser scanners

Laser scanners are widely found in space related pose estimation methods [3, 37, 80]. The large advantage is the combination of an active sensor (and thus, the resulting robustness against harsh lighting conditions) with the sensitivity of the receiver optimized for the wavelength used. However, care must be taken during the design of the laser scanner mechanics, otherwise the launch of the spacecraft can lead to irreversible damage. Also, the update rate can be limited, as laser scanners have severe scanning speed limits.

Laser scanners provide point clouds of the target object along with brightness information for each point. The most common algorithm used to align known models with the point cloud is the Iterative Closest Point algorithm [120]. ICP minimizes the squared distances between nearest points in the point cloud and the model, while optimizing the pose estimate. It requires a rough alignment of the point clouds in order to work, because it can not find the global minimum of the error metric. Hence, approaches of using the state estimate of a Kalman filter as a rough alignment are widely used [1].

The Principal Component Analysis is also often used to help with the initial alignment of the model [63]. Once a pose estimate is available, an ICP variant continuously tracks the target and updates the pose estimate.

In essence, the large advantage of laser scanners is their range (when using laser scanners based on the time-of-flight principle) and independence on the environmental lighting conditions. However, the drawback is high cost, sensitive mechanics and – compared to CCD or PMD sensors – higher power requirements.

8.1.4 Stereo vision

Stereo vision is very common in Rendezvous and Docking, as the sensors are inexpensive and the resulting 3D point clouds facilitate pose estimation. However, the drawback is often the limited field of view [14, Chapter 7] or the extensive processing time required. For example in [138], the stereo matching takes about 5 seconds, and the iterative closest point method takes another 5-11 seconds for a single image. This is likely a consequence of the high number of points in the model and the point cloud, which must be matched.

Robustly tracking images using stereo vision and by using the disparity information, gathering distance information, is pursued in [100]. While the method can cope with the loss of a camera, once the algorithm was initialized (stereo disparity information is not required), the initialization becomes weaker and less accurate. The target object may be unknown and is tracked using image features. Rotation errors are within 5°, translational errors are within 6 cm. The most prominent advantage of the approach is that the method works without object information.

When a model of the object is available, the pose may be estimated by iteratively refining a fit based on rendered model edges and detected edges in the image [62, 69]. However, rendering the entire model can be computationally too demanding for embedded systems, because there is no dedicated graphics hardware available. Alternatively, the 3D model can be directly fitted into the point cloud.

Finally, a comparison of monocular pose estimation and stereo pose estimation is made in [8]. In essence, the most important result of the comparison is that stereo vision is better for estimating the distance of objects at large distances and under skew of the target object.

8.1.5 Summary and overall performance comparison

In general, comparing approaches of different architecture is very difficult. There is no common setup and the objects used range from scaled satellites to original size mockups and in some cases, other objects. Still, when comparing the raw pose deviations, at least a rough impression is gained. Table 8.1 shows the results in terms of accuracy of several selected papers, which are compared to the method developed here.

The accuracy numbers in the table are roughly what can be expected as the worst case error for the individual method, when neglecting situation-specific outliers. Not all authors provide mean errors and standard deviations, what is why this has been chosen to be the quantity used for the comparison. The values have been extracted from plots in the relevant papers when it was necessary, as described in the previous subsections. For the statistical error analysis related to this work, see table 7.5 on page 172. It has to be kept in mind here that the degree of realism in the simulation/image generation differs widely and it can be safely said at this point, that the work

Table 8.1. Overall performance comparison of different model-based pose estimation methods, using different sensors.

Sensor(s)	Method	Reference	Angular accuracy* (deg)	Positioning accuracy*(m)
1× CCD	Features, closed-form pose est.	Philip and Ananthasayanam [104]	0.3	0.001
1× CCD	Features, closed-form pose est.	Zhang and Cao [154]	1.2	0.25
1× CCD	Monocular cues[†]	Barrois and Wöhler [9]	0.8	0.004
2× CCD	Features, dual-quaternion optimization	Oumer and Panin [100]	5.0	0.06
2× CCD	Model Edge tracking	Kelsey et al. [69]	3.3	0.021
LiDAR	ICP	Jasiobedzki et al. [63]	2.0[‡]	0.05[‡]
PMD, CCD	Fused edge/ surface tracking	Figures 7.10, 7.12, 7.13, 7.14	3.5	0.04

[*] Accuracy here means *worst case error* in order to facilitate the comparison, but situation-specific outliers have been neglected. Note that differences in the test scenarios, which have an influence on the results, can not be compensated.
[†] Limited distance only. Used cues: defocus, brightness.
[‡] Achieved with point clouds of 2,000 points or more.

at hand has used the most sophisticated simulation of them all (i.e., less realistic image material will very likely lead to better results).

Apparently, purely CCD based approaches can be very accurate. However, it is more likely that the errors remain small, as the distance remains small. The fact that a single pixel gains impact with increasing distance can not be ameliorated by any method. It is interesting to see that stereo based pose estimation is approximately as good as single-CCD approaches, but there is clearly more potential in large-distance scenarios.

The laser scanner example is not representative, as there are much more expensive and sophisticated scanners available. However, comparing systems which have a financial value factor of 1,000 or above between them, does not make much sense.

Figure 8.1. The Kinect sensor. An infrared emitter projects a pattern into the scene, which is captured by a camera. Similar to stereo vision, the disparity is used to obtain the distance measurement. (*Image from Wikipedia, public domain*)

Being a bit out of line with the approach presented within this thesis, it is still nice to see that this fused approach can definitely compare to established methods.

8.1.6 Kinect sensor

Before moving on to discussing the impact of environmental conditions on the PMD sensor, for the sake of completeness, a few words about the Kinect sensor seem to be appropriate: In a review of a paper, a reviewer rose the question of how the PMD sensor is related to the Kinect sensor developed by Microsoft and PrimeSense. This is why at this point, a brief discussion about the differences between the Kinect sensor shown in figure 8.1 and the PMD sensor is made.

The Kinect sensor works by emitting a pseudo random pattern and projecting the pattern onto the scene using an infrared light emitter. A CCD sensor with sufficient sensitivity in the infrared spectrum then captures the scene with the projected pattern. By using stereo vision techniques, the distance of a point in the image can be recovered by measuring the disparity.

While stereo vision methods must rely on texture and estimate the disparity by matching image pixels from two images, the Kinect sensor relies on the pattern, what means this will work particularly well for untextured surfaces, where traditional stereo matching is likely to fail.

The primary motivation for this comparison is the question whether the Kinect sensor is as well suited for space as the PMD camera, as it is clearly similar from a functional point of view, but achieves the goal with a completely different approach. A large problem is that the Kinect sensor is not very well investigated by the scientific community yet. However, performance data is available [5] and a brief comparison is provided in table 8.2.

What can be seen at first glance is that the Kinect sensor has a much higher spatial resolution. Also from the frame rate, the Kinect sensor looks promising. However, the first problem is the distance range. While the PMD camera can theoretically

Table 8.2. Comparison of the PMD sensor and the Kinect sensor. The Kinect sensor data was taken from [5].

Property	PMD camera	Kinect sensor	Unit
Framerate	< 20*	≈ 30	Hz
Spatial resolution	200 × 200	640 × 480	pixels
Depth resolution	< 0.01	0.01	m
Minimum distance	0.1	0.8	m
Maximum distance	7.5†	3.5	m

* Depending on the chosen integration time, higher imaging rates may be possible. For very high integration times, the frame rate can drop significantly.
† up to 75 m with multi-frequency imaging, when using 18 and 20 MHz as modulation frequencies, as suggested in section 4.2 on page 105.

measure distances of up to 75 meters with the dual-frequency imaging approach presented in this thesis, the Kinect sensor is limited to 3.5 meters, what is nowhere near the PMD camera. Moreover, this is a disadvantage which can not be significantly ameliorated by designing a different pattern or enhancing the resolution of the sensor. Instead, this is an inherent limitation of the working principle.

Second, one has to keep in mind that the target object may be highly articulated and especially MLI shielding poses a large problem for any active sensor relying on reflection of a signal. In this case, the Kinect sensor can be expected to run into problems due to heavy distortion of the emitted infrared pattern. Having to rely on a pattern instead of just reflected light makes the process even more sensitive.

Finally, the PMD camera was designed to work in sunlight and more importantly, outdoors. In contrast, the Kinect sensor was designed for indoor applications and it can be safely said that it will be certainly much more sensitive to sunlight than the PMD camera.

In essence, it seems that the Kinect sensor has several advantages (frame rate, spatial resolution) and is well suited for gesture recognition. However, it is currently far from being of use in a space environment and the architecture does not allow that to change much in the future (mainly due to the inherent limitation in the distance range). It is therefore concluded that the PMD camera is technologically superior with regard to the application targeted by this thesis, despite its own drawbacks.

8.2 PMD sensor space suitability considerations

So far, the pose estimation method has been discussed under various conditions, however, there is another problem which is related to the PMD sensor. As it has not been qualified and tested for use in a space environment, it remains to be seen whether state-of-the-art sensors are usable at all or if hardware modifications are necessary. In this section, the most important space-related environmental conditions are briefly discussed with regard to the PMD sensor.

8.2.1 Thermal

Depending on the local temperature, there will be a different number of thermally generated electrons (dark current). This has an impact on the measurement [127]. Any attempt of calibrating the sensor for a specific temperature, as it is currently being done by all approaches published so far (including this document) assume that the sensor is used in thermal equilibrium. This is not possible in a space environment. The temperature variations can be reduced by cooling and heating systems, but these are utmost expensive, as they require large amounts of energy and there are still limits, which can not be overcome.

As a consequence, the effect must be accepted to be present and compensation measures must be implemented which can deal with the symptoms of the effect, instead of trying to remove it completely by eliminating its causes. The obvious method would be to implement a temperature sensor as close to the sensor chip as possible (the best solution would be an on-chip thermal diode). Then, when the temperature is known, the effect may be compensated during the processing of the sensor data.

Another distinction must be made. The necessity of calibrating this effect emerges from the orbit parameters of the spacecraft in question. When the orbit parameters allow for continuous sunlight or short to zero eclipse periods (such as some special cases of sun-synchronous orbits), a thermal equilibrium can be established that prevents the necessity of calibrating the effect. However, for most applications and orbits, different eclipse phases will occur [26] and the only way of compensating for the effect is the above suggested temperature measurement and further processing of the sensor data in the preprocessing stage. Otherwise, there will be periods of several minutes up to tens of minutes, where the camera measurements are continuously drifting as a consequence of heat exchange and temperature change.

8.2.2 Radiation

In general, there is currently no data available on the effects of radiation on the PMD chip. A comparison with the typical effects of CMOS CCD imagers is also not envisaged here, as PMD sensors require a different working principle for the determination of the distance. Still, as one of the most important effects caused by radiation in the case of CMOS imaging sensors is the increase of dark current [49], it can be speculated that PMD sensors will experience an increase of the non-linear effects of single pixels.

Since the potential offsets do not apply to all elementary cells at once, the result will be an almost impossible-to-predict deformation of the distance estimation. However, the magnitude of the effect may be minimal; a statement is not possible without precise knowledge about the sensor.

When it comes to compensation measures, detecting these deformations will be utmost difficult, if not impossible. The only measure which may be applied in practice is the detection of pixels damaged beyond negligible offsets. Such pixels may be marked as being defect and would be excluded in future measurements. As most PMD data processing algorithms work with point clouds, the result will be a sparsified point cloud, which would allow for graceful degradation.

Such pixels may be detected by investigating the values of the elementary cells over time (value range checks, comparison with neighbor pixels, etc). Also, consistency checks (as proposed in section 2.4.2 on page 58) are viable.

8.2.3 Launch vibration

Once a satellite has reached its orbit, there are no significant vibrations, which may have an influence on the performance of a sensor system such as the PMD camera. However, during the launch, strong forces and vibrations can damage sensitive systems or permanently influence their properties [83], which is why all systems designed for space applications undergo vibration tests.

When it comes to the PMD camera, especially the housing of the PMD camera and the interface to the optics is more or less likely to be damaged as a consequence of vibrations. Electronics usually fail depending on the shape of the individual components, the PCB thickness, and – most importantly – the mechanical interconnections of the PCB and the housing. The local acceleration itself can not be used as an indicator for failure [4].

Typical points of failure which must be mentioned here as well, are all sorts of electrical connectors which is why it is recommended to use connectors capable of storing mechanical energy up to a certain point [136]. Connectors not only fail due to vibration but also because of thermal stress, especially when the compounds used have different coefficients for temperature related expansion. The referenced paper suggests that well-designed elasticity near connectors can ameliorate this problem. As the external connectors of the PMD camera are threaded, it can be assumed that these connections are rather stable in this regard.

8.3 Conclusions

8.3.1 Summary

In this thesis, a method was presented that allows to use a PMD sensor in conjunction with a CCD sensor on board a spacecraft for estimating the relative pose of a neighboring satellite in a formation or a satellite target during a rendezvous maneuver.

A range extension of the PMD sensor was presented, which can be used specifically in a space environment. Previous approaches have relied on the geometry of the scene, what becomes impractical, once there are no walls or ground in the image. Moreover, for the first time, a PMD sensor could be used for long range measurements on a calibrated test facility. Also, the sensor calibration must be designed with respect to multi-frequency measurements. It was shown that by designing compensation methods specifically for using them in periodic cases, the necessary PMD camera data preprocessing can be suitably adapted to provide high accuracy long-range measurements.

As stated in the Kinect sensor comparison, it is unlikely that such a range extension is possible with the Kinect sensor, as its range is bound by the resolution of the CCD sensor imaging the pattern. Therefore, the PMD sensor is the right choice. It can also be assumed true that current implementations of the Kinect sensor do not have any means of filtering environmental infrared light, in contrast to the PMD camera.

This is specifically important for the target acquisition method. The robustness of this part of the algorithm comes primarily from the measurement principle of the PMD camera, but also relies on the specific properties of applications in space. In contrast to previous works, the resolution of object symmetries is addressed with a framework that allows to pass additional information to the algorithm. It should also be pointed out that the target acquisition takes less than one second on recent computer hardware, and that is without any form of optimization.

The approach of fusing CCD and PMD sensor data by combining deviation components extracted from the images provides a new and computationally very efficient way of using multiple sensors for pose estimation, what outperforms methods using only the PMD sensor, and also outperforms most CCD-based methods as soon as the object distance increases. It also outperforms most laser-scanner sensors in terms of cost. The high frame rate of the sensors propagates through the processing chain and allows measurements even for objects moving or rotating at higher velocities, what is in particular useful for the case of tumbling target objects.

For the first time, measurements with a CCD- and PMD-sensor have been made in the presence of a light source that is very close to the real sun spectrum in orbit. Compared to previous approaches, this allows an evaluation of the sensors to be much closer to the real conditions. When the environment does not provide sufficient light for the CCD sensor to operate, tests have shown that a method purely based on the PMD measurements, such as the initializer proposed in this thesis, are capable of providing a pose estimate, although at a reduced accuracy. Furthermore, tests with simulated sunlight have shown that there are as well situations, where the PMD sensor can not be used due to light which is too strong and saturates the PMD cells, while the CCD sensor can still be used. Here, pose estimation techniques purely based on CCD sensors could provide measurements. Another conclusion from the tests is that overall, the PMD sensor is – in its current stage – not suitable for operation in space due to the sensitivity to environmental light. However, a laser-based illumination system along with an optimized optical filter path looks very promising in that it would allow the developed algorithms to be used without modification while at the same time providing more sunlight tolerance, thereby widening the field of application.

8.3.2 Algorithm limits

While the sensors impose limits on the environmental lighting, there are also limits related to the situation, of which the most important ones are now briefly discussed. For suggested solutions to these limits, see the next section.

Visibility of the target object

Since the presented algorithm relies on tracking the edges of the target object with the CCD camera, there is a constraint that these edges must be visible at all times. More importantly, the target must not move too close to the chaser. Otherwise, the outer edges will leave the image space. To counteract this problem, a tracking algorithm based on images features can provide measurements while the distance is reduced until the docking process has been completed.

Size of the target object

Due to perspective projection the object will appear smaller, the larger the distance between the chaser and the object grows. There is a point where an accurate pose estimation is no longer possible. This was not yet relevant for the measurement campaign performed within this thesis, however, it is important to keep that in mind. This does, of course, also depend on the target size. With the mockup used here, one could estimate that this point would possibly be somewhere around 30–40 meters.

High-frequency structures

MLI and instruments are not a problem, as long as their influence is limited. Nevertheless, the approach of estimating the pose of an object presented in this thesis is based on articulated edges and surfaces of an object. As soon as the object does not have well articulated edges or surfaces, this method can not be applied any longer.

Therefore, there is a strong limitation on the outer shape of the target object. However, please note that especially the planarity constraint is a consequence of the low lateral resolution of the PMD sensor, which can not capture high-frequency features very well.

8.4 Future work

Regarding the presented pose estimation algorithm, there are a few research topics which could improve the results or solve remaining problems. Such ideas will be briefly discussed in the following.

8.4.1 Feature-based docking tracker

As stated in section 8.3.2, once the target is very close to the chaser, it may happen that one or more of the outer edges of the target object are not visible any more because they have moved outside the image area. When this happens, the presented algorithm can no longer reliably determine the pose. Therefore, an on board computer may switch to a different algorithm for retrieving pose estimates.

Here, it is suggested that a feature-based algorithm is used. The advantage of feature-based algorithms is that tracking is possible as long as at least three features are available (for tracking in three dimensions). Such an algorithm was already developed earlier [143] (what allows tracking as soon as two features are available). The main problem here was, however, the low image-space resolution of the PMD camera. With the CCD camera available, this method could be extended in such a way that brightness information is extracted from the high-resolution CCD image and distance information (for estimating the normal vector of the feature) is extracted from the PMD camera.

8.4.2 Extended range

While it was possible to extend the measurement range of the PMD camera, so it is sufficient to test it inside the EPOS facility, it would also be of interest to see how the algorithm performs at even larger distances. This thought is primarily motivated from the effect, that the target will become smaller with increasing distance. As a result, there will be a point, where the proposed method will no longer work. The location of this point could not be determined within this thesis (a very rough estimation was given in section 8.3.2 on page 203). Therefore, performing outdoor tests would very likely answer this question.

When this critical distance is reached, it is very likely that six degrees of freedom are not of primary interest, but the direction and the distance. This is a problem which has already been partially solved within a PRISMA experiment [28]. This experiment did not have a PMD camera available and the distance was not estimated, but the direction was estimated. A different approach was made earlier in the predecessor of the method used for the PRISMA experiment, and here the distance was estimated as well by relying on the size of the object [141]. Assuming there would be a modulated laser available, and using this as a light source, distances of 75 meters and maybe even more (with some effort, for example, using more than two frequencies) would become measurable with the PMD camera. Alternatively, the modulation frequencies of the camera can be tuned. Lower frequencies allow for larger measurement ranges [14, page 114] (see also section 4.2.3 on page 112).

Even if only one pixel of the target could be measured, the distance would not have to be inferred from the size of the object in image-space. The result would be a large gain in distance estimation accuracy.

8.4.3 Hidden object removal

While the tracking of the target provides good results as long as there is no partial occlusion, this can turn into a problem for a target with an articulated surface. For example, the apogee thruster could very easily occlude one of the outer borders, when rotated far enough. The current implementation has two ways of proceeding with pose estimation despite this problem: First, the tracker will switch and track a different surface of the object, as soon as the angular measurement of the normal vector of the tracked surface exceeds a threshold. Second, partially occluded edges do not pose a problem, as long as other edges are visible and the edge in question does not produce wrong tracking input.

Still, at this point it would be of use to have a means of removing occluded model edges. This has been done already numerous times in computer graphics. The reason why this has not yet been implemented was a lack of time and the initial requirement to be independent of computer graphics libraries. On the real-time system used for the tests, there is no computer graphics interface available and porting libraries such as MESA[3] to VxWorks is prohibitively time-consuming at best. In essence, this can be seen as a low-priority improvement.

8.4.4 Laser illumination

When using a laser instead of infrared LEDs, the optical filter path can become more narrow in terms of bandwidth. Due to the fact that LEDs have a rather broad wavelength range, the filter must also be designed to allow a broad range of light to pass through unless one can afford signal dampening. Lasers have line spectra and therefore qualify to solve the problem. Once the band pass for the active light in the PMD camera can be narrowed down, there will be also less environmental light, which interferes with the measurement. Consequently, the PMD camera can be tuned even more to withstand strong sunlight influence.

The necessary hardware changes would encompass providing an infrared laser which can be modulated at frequencies of about 20 MHz with a wavelength near the specified LED wavelength (830 – 870 nm), as well as a newly designed optical filter path, which has its central transmission wavelength coinciding with the one of the laser, and a very narrow transmission window around it.

8.4.5 Initializer object symmetry resolution

As stated in section 5.2.5 on page 127, the current implementation of the initializer can not reliably determine all six degrees of freedom of an object, when the object is symmetric. Despite the already available methods for providing hints to the initializer, which allow overcoming this problem in most cases, it would be advantageous to remove the need for these hints.

One possible solution could be based on the feature-based pose estimation method developed earlier [143]. In this case, it would be sufficient to detect one feature. When the feature orientation information is not used or not available (what would be the case when using features such as SIFT [93] or SURF [11], as both do not

---

[3]The MESA 3D graphics library is a software OpenGL-backend. It can be found at http://www.mesa3d.org/

account for 3D rotation), two matched features would suffice. The features should be extracted from the CCD image due to the higher resolution of the pixel grid.

8.4.6 Environmental effects

The investigation of environmental effects, such as temperature changes, radiation and air pressure changes are out of scope of this thesis. Consequently, the effects these environmental conditions might have on the PMD sensor have not been evaluated. An investigation of these effects should answer two questions: First, is the respective effect significant enough to distort the measurement in a destructive way and second, if yes, can it be compensated using a suitable calibration or can it be filtered? If nothing can be done about the effect in software, what are suitable hardware modifications to achieve the same goal?

8.5 Outlook

Clearly, autonomous Rendezvous and Docking is of great interest. All nations known for their space experience have programs running for the development of autonomous Rendezvous and Docking capabilities (the efforts have been summarized in section 1.2 starting on page 23). Even after the financial crisis of 2008, where research funding has substantially decreased, the topic remains of utmost importance. From a different point of view, On-Orbit Servicing is a very interesting, active research area, which will bring a lot of new knowledge in the near future.

Thinking of DEOS and other projects, space debris and the ever increasing need for exploration – the potential and the drivers are there. The space debris problem is of high importance, when regarding the wish to operate satellites in the future without a lot of additional effort. During the time this was written and while working at DLR, several evasive maneuvers had to be made with satellites, which are being controlled by DLR, due to space debris and collision warnings.

From prevention of creating new space debris to collecting it, autonomous Rendezvous and Docking, as well as autonomous object tracking are (and will remain) important fields for research. The PMD sensor and especially promising developments of combined RGB/ depth sensing chips [71] show that this technology is still growing to become more important over time and also indicates that PMD sensor technology continues to evolve. It can be anticipated, that the first experiments with PMD cameras in space will materialize in the next few years.

List of Figures

1.1	Space debris	19
1.2	Algorithm flow chart (preprocessing part)	32
1.3	Algorithm flow chart (principal part)	33
2.1	The old EPOS facility	36
2.2	The new EPOS facility	38
2.3	The structure of EPOS	39
2.4	Perspective projection and image-space coordinates	42
2.5	Calibration of the CCD camera	47
2.6	Optical distortion of the CCD camera	50
2.7	PMDTec Camcube 3.0	51
2.8	Architecture of a PMD pixel (single channel)	53
2.9	Visualization of the quantities measured by the PMD camera	55
2.10	Principle of SBI	57
2.11	Optical distortion of the PMD camera	60
2.12	Coordinate frames and -transforms	61
2.13	Location and orientation of sensor coordinate frames	64
2.14	The result of the texture segmentation process	68
3.1	Quantities measured by the PMD camera	76
3.2	Quantities measured by the CCD camera	77
3.3	OpenGL visualization of the OLEV model	79
3.4	Model preprocessing	81
3.5	Line parsing	85
3.6	Triangle processing	86
4.1	Visualization of fixed pattern noise	97
4.2	Wiggling effect calibration (20 MHz)	98

4.3	Wiggling effect calibration (18 MHz)	99
4.4	Fit of the Wiggling effect error model	100
4.5	Integration time offset for 20 MHz	101
4.6	Integration time offset for 18 MHz	102
4.7	PMD Amplitude offset error compensation (20 MHz)	103
4.8	PMD Amplitude offset error compensation (18 MHz)	104
4.9	Multi-frequency visualization	108
4.10	Distance difference plot for two frequencies	109
4.11	Restored distance from two simulated PMD measurements	111
4.12	DDAE Range/ Noise trade-off	113
5.1	Pose estimator state diagram	118
5.2	Initializer structure	119
5.3	Target acquisition visualization	119
5.4	Determination of principal components of a point cloud	124
5.5	Initial attitude quaternion determination	126
6.1	Different approaches for sensor data fusion	130
6.2	Principle of the scanline algorithm	135
6.3	Scanline visualization	138
6.4	Image space deviation components	141
6.5	Roll angle residual visualization	142
7.1	EPOS mounting plate & sensor placement	151
7.2	Satellite mockup	152
7.3	Comparison of real and simulated sun spectrum	154
7.4	Prosilica GC-655 spectral response	154
7.5	Image capturing component overview and data flow	157
7.6	PMD modulation frequency measurements	159
7.7	PMD modulation signal timing measurements	160
7.8	Pose estimation visualization	164
7.9	Visualization of the distance measurement	165
7.10	Position estimation error for the linear approach	166
7.11	Wiggling effect residuals after compensation	166
7.12	Angular estimation error for the linear approach	167
7.13	Compound motion translational error	169

7.14 Compound motion angular errors 170
7.15 Compound motion amplitude 170
7.16 Target affinity matrices for different distances 174
7.17 Target acquisition analysis (position) 175
7.18 Target acquisition analysis (orientation) 176
7.19 Initializer image amplitude scenarios 177
7.20 Initializer state variable progression 177
7.21 Tracker convergence after bad initialization 179
7.22 Tracker convergence, position error 180
7.23 Tracker convergence, orientation error 181
7.24 Erroneous target acquisition recovery 182
7.25 Rear sun illumination (CCD sensor) 183
7.26 Rear sun illumination (PMD sensor) 184
7.27 Side sun illumination 186
7.28 Tracking performance under sunlight 187
7.29 Frontal sun illumination (CCD sensor) 187
7.30 Frontal sun illumination (PMD sensor) 188

8.1 The Kinect sensor 197

List of Tables

2.1	Comparision of EPOS facility performance indices	40
2.2	Intrinsic calibration results for the monocular camera	48
2.3	PMDtec CamCube 3 Specifications	52
2.4	Intrinsic calibration results for the PMD camera	59
2.5	Extrinsic sensor parameters	65
3.1	Search space separation	78
4.1	Chosen modulation frequencies and practical consequences	112
7.1	Comparison of absolute irradiance levels relevant to the PMD sensor	155
7.2	Sensor latency statistics	158
7.3	Properties of the linear approach experiment (ideal conditions)	162
7.4	Comparison of pose estimation errors	171
7.5	Static pose estimation error statistics	172
7.6	Static target acquisition error statistics	174
8.1	Overall performance comparison	196
8.2	Comparison of the PMD sensor and the Kinect sensor	198

List of Algorithms

3.1	Surface finding algorithm	80
3.2	Triangle processing algorithm	82
3.3	Edge processing subroutine (addEdge)	82
3.4	Vertex/ outer edge mapping	83
3.5	Line extraction	83
3.6	Line extraction subroutine (parseLine)	84
4.1	Normalized coordinate estimation	96
5.1	Procedure for determining the surface affinity matrix	121
6.1	Determination of the dominant surface	135

Nomenclature

ACPI   Advanced Configuration and Power Interface, a standard defining interfaces to power management and in general, the hardware, on today's computer systems.

ACS   Application Control System, a computer system of the EPOS facility designed to operate the Simulink models.

ARGON   Advanced Rendezvous experiment using GPS and Optical Navigation, an experiment carried out using the PRISMA satellites, where an optical far range navigation algorithm was demonstrated.

ATV   Autonomous Transport Vehicle, a container for transporting goods to the ISS.

AVI   Audio/ Video Interleaved, a container file format specification for video data.

CAM   Computer Aided Manufacturing, a combination of software and hardware which accelerates and simplifies the production of real-world items.

CCD   Charge Coupled Device, a vision sensor architecture based on the photoelectric effect.

CCFL   Cold-Cathode Flourescent Lamp, a lamp type using ionized gas as source of light.

Chaser   The active satellite in a Rendezvous and Docking scenario, which initiates all procedures and uses its actuators to complete the process.

DDAE	**D**irect **D**istance **A**mbiguity **E**stimation, a method for extending the measurement range of PMD sensors by using multiple modulation frequencies.
DEOS	**DE**utsche **O**rbitale **S**ervicing Mission, a German Rendezvous and Docking demonstration mission planned for 2017 (or later).
DLR	**D**eutsches Zentrum für **L**uft- und **R**aumfahrt (German Aerospace Center)
DoF	**D**egree **of** **F**reedom
EPOS	**E**uropean **P**roximity **O**perations **S**imulator, a facility for testing Rendezvous and Docking operations and sensors.
ESA	**E**uropean **S**pace **A**gency
EtherCAT	A real-time capable bus system based on Ethernet.
FMC	**F**acility **M**onitoring and **C**ontrol, the primary control system of the EPOS facility.
FPGA	**F**ield-**p**rogrammable **G**ate **A**rray, an integrated circuit consisting of logic cells which can be interconnected to form functions.
GEO	**G**eostationary **E**arth **O**rbit, orbits grouped by altitude of 35786 km.
GPU	**G**raphics **P**rocessing **U**nit, a processor designed specifically for calculations related to graphics processing, mainly rendering. GPUs have typically massively parallelized, pipelined floating point units.
HEO	**H**igh **E**arth **O**rbit, orbits with an altitude larger than the GEO altitude.
ICP	**I**terative **C**losest **P**oint, a method of iteratively aligning two point sets of an object (in most cases, model and measurement). Commonly found at the core of 3D pose estimation algorithms.
ISS	**I**nternational **S**pace **S**tation

Kinect	A real-time depth-imaging sensor developed by Microsoft, targeted mainly for consumer applications.
LED	<u>L</u>ight <u>E</u>mitting <u>D</u>iode, a semiconductor structure optimized for light emission.
LEO	<u>L</u>ow <u>E</u>arth <u>O</u>rbit, orbits grouped by altitude in the range of 80 – 2000 km.
LEOP	<u>L</u>aunch and <u>E</u>arly <u>O</u>rbit <u>P</u>hase, a slice of time dedicated to launching a satellite and bringing it into routine operations.
LiDAR	<u>Li</u>ght <u>D</u>etection <u>A</u>nd <u>R</u>anging, an active sensor measuring the distance to objects by the Time-Of-Flight principle.
LSH	<u>L</u>ocality <u>S</u>ensitive <u>H</u>ashing, a method to determine parameters from inputs by exploiting collisions of suited hash functions.
LVDS	<u>L</u>ow <u>V</u>oltage <u>D</u>ifferential <u>S</u>ignaling, a method for transporting high-frequency digital signals robust to external interference, while causing as little interference as possible by itself.
MEO	<u>M</u>iddle <u>E</u>arth <u>O</u>rbit, orbits grouped by altitude in the range of 2000 – 35786 km.
MLI	<u>M</u>ulti <u>L</u>ayer <u>I</u>nsulation, a thin foil composed of metal coated plastic layers to stabilize the temperature inside satellites by dampening heat dissipation in both directions.
MMI	<u>M</u>an-<u>M</u>achine <u>I</u>nterface, computer systems designed for interaction with EPOS operators.
OLEV	<u>O</u>rbital <u>L</u>ifetime <u>E</u>xtension <u>V</u>ehicle, a satellite designed to dock to GEO satellites and thereby extend their life.
PCA	<u>P</u>rincipal <u>C</u>omponent <u>A</u>nalysis, a method for finding base vectors along a data set, such that the variance of the data set among the new base vectors is maximized.

PCB	<u>P</u>rinted <u>C</u>ircuit <u>B</u>oard, the union of substrate material, copper layers and soldered electronic components.
PM	<u>P</u>ower <u>M</u>anagement, the ability of an embedded system, hardware component or computer system to reduce its own power consumption by slowing down clocks (throttling), partially stopping clocks (clock gating) and cutting power to unneeded hardware parts.
PMD	<u>P</u>hotonic <u>M</u>ixer <u>D</u>evice, a sensor architecture capable of determining reflectivity and distance based on phase measurement.
Pose	The union of position (spatial location, represented by a vector) and orientation of an object (represented by rotation matrices, quaternions or other characterizations).
PRISMA	<u>P</u>rototype <u>R</u>esearch <u>I</u>nstruments and <u>S</u>pace <u>M</u>ission technology <u>A</u>dvancement, a pair of satellites and also the name of a swedish mission for demonstration of formation flight in space.
RT	<u>R</u>eal-<u>T</u>ime, a short term used for computer systems of the EPOS facility running real-time operating systems, such as VxWorks and RTLinux.
SBI	<u>S</u>uppression of <u>B</u>ackground <u>I</u>llumination, a method for reducing the influence of background illumination on PMD sensor measurements.
SimCon	<u>Sim</u>ulation <u>Con</u>nection, a MATLAB/ Simulink block for connecting Simulink models across multiple computers.
SLAM	<u>S</u>imultaneous <u>L</u>ocalization <u>a</u>nd <u>M</u>apping, a keyphrase for the self-pose estimation problem and environment mapping- commonly found in mobile robot research and corresponding literature.
SNR	<u>S</u>ignal to <u>N</u>oise <u>R</u>atio, a relative measure of signal quality.
SVD	<u>S</u>ingular <u>V</u>alue <u>D</u>ecomposition, a method for splitting a matrix **M** into two unitary matrices **U**, **V*** and a diagonal matrix Σ containing the singular values of **M**.

Target	The uncooperative satellite in a Rendezvous and Docking scenario, which will remain passive with regard to actuators.
TCP	Transmission Control Protocol, a connection oriented, reliable inter-system communication protocol, which is part of the internet protocol family.
TLC	Target Language Compiler, a language for platform-independent code generation used by the Simulink Real-time workshop.
ToF	Time of Flight, a principle of measuring the time a signal takes to reach a destination and returns, and thus, by assuming a constant velocity, determining the distance.
VME	Versa Module Eurocard, a parallel bus system.

Bibliography

[1] F. AGHILI, M. KURYLLO, G. OKOUNEVA, and D. MCTAVISH. "Robust Pose Estimation of Moving Objects Using Laser Camera Data for Autonomous Rendezvous and Docking." In: *Proceedings of the ISPRS Workshop on Laserscanning*. Volume XXXVIII.3/W8. 2009, pages 253–258 (cited on page 194).

[2] M. ALBRECHT. "Untersuchung von Photogate-PMD-Sensoren hinsichtlich qualifizierender Charakterisierungsparameter und -methoden." Dissertation. University of Siegen, Germany, 2007 (cited on page 93).

[3] A. C. ALLEN, N. N. MAK, and C. S. LANGLEY. "Development of a Scaled Ground Testbed for Lidar-based Pose Estimation." In: *Proceedings of the IEEE/RSJ International Conference on Intelligent Robots and Systems-Robot Vision for Space Applications (RVSA)*. 2005, pages 10–15 (cited on page 194).

[4] R. A. AMY, G. AGLIETTI, and G. RICHARDSON. "Board-Level Vibration Failure Criteria for Printed Circuit Assemblies: An Experimental Approach." In: *IEEE Transactions on Electronics Packaging Manufacturing* 33.4 (2010), pages 303–311. ISSN: 1521-334X. DOI: 10.1109/TEPM.2010.2084092 (cited on page 201).

[5] M. ANDERSEN, T. JENSEN, P. LISOUSKI, A. MORTENSEN, M. HANSEN, T. GREGERSEN, and P. AHRENDT. *Kinect Depth Sensor Evaluation for Computer Vision Applications*. Technical report ECE-TR-6. Denmark: Aarhus University, 2012 (cited on pages 197, 198).

[6] A. ANDONI and P. INDYK. "Near-optimal hashing algorithms for approximate nearest neighbor in high dimensions." In: *Communications of the ACM* 51.1 (Jan. 2008), pages 117–122. ISSN: 0001-0782. DOI: 10.1145/1327452↩.1327494 (cited on page 116).

[7] J. APELDOORN, R. JANOVSKY, S. MERIT, R. DESTEFANIS, F. SCHAEFER, J. GELHAUS, and E. IMRE. "Improvements for space mission protection against space-debris." In: *Proceedings of the 62nd International Astronautical Congress*. Cape Town, ZA, 2011 (cited on page 20).

[8] P. AZAD, T. ASFOUR, and R. DILLMANN. "Stereo-Based vs. Monocular 6-DoF Pose Estimation Using Point Features: A Quantitative Comparison." In: *Autonome Mobile Systeme 2009*. Informatik aktuell. Springer Berlin Heidelberg, 2009, pages 41–48. ISBN: 978-3-642-10283-7. DOI: 10.10↩ 07/978-3-642-10284-4_6 (cited on page 195).

[9] B. BARROIS and C. WÖHLER. "3D Pose Estimation Based on Multiple Monocular Cues." In: *Conference on Computer Vision and Pattern Recognition*. IEEE Computer Society, 2007. DOI: 10.1109/CVPR.2007.383352 (cited on pages 193, 196).

[10] N. BAUER. *Handbuch zur industriellen Bildverarbeitung: Qualitätssicherung in der Praxis*. Stuttgart: Fraunhofer-IRB-Verlag, 2007. ISBN: 978-3-81-677386-3 (cited on page 123).

[11] H. BAY, T. TUYTELAARS, and L. V. GOOL. "SURF: Speeded Up Robust Features." In: *9th European Conference on Computer Vision*. Graz Austria, May 7–13, 2006. DOI: 10.1007/11744023_32 (cited on pages 133, 206).

[12] H. BENNINGHOFF, T. BOGE, and T. TZSCHICHHOLZ. "Hardware-in-the-loop rendezvous simulation involving an autonomous guidance, navigation and control system." In: *Proceedings of the 1st IAA Conference on Dynamics and Control of Space Systems*. 2012 (cited on page 163).

[13] R. BEVILACQUA, M. ROMANO, and O. YAKIMENKO. "Online generation of quasi-optimal spacecraft rendezvous trajectories." In: *Acta Astronautica* 64.2-3 (2009), pages 345–358. ISSN: 0094-5765. DOI: 10.1016/j.ac↩ taastro.2008.08.001 (cited on pages 75, 93).

[14] D. A. BHATTI, editor. *Stereo vision*. InTech, Vienna, Austria, 2008. ISBN: 978-953-7619-22-0. DOI: 10.5772/89 (cited on pages 51, 107, 112, 194, 205).

[15] J. BIOUCAS-DIAS and G. VALADAO. "Phase Unwrapping via Graph Cuts." In: *IEEE Transactions on Image Processing* 16.3 (2007), pages 698–709. ISSN: 1057-7149. DOI: `10.1109/TIP.2006.888351` (cited on page 106).

[16] T. BOGE. *EPOS customer document*. EPOS/DLR/ECD-5/01-2.00. 2003 (cited on page 40).

[17] T. BOGE, H. BENNINGHOFF, M. ZEBENAY, and F. REMS. "Using robots for advanced rendezvous and docking simulation." In: *ESA Workshop on Simulation and EGSE facilities for space programmes* (2012) (cited on page 156).

[18] T. BOGE, T. RUPP, K. LANDZETTEL, T. WIMMER, C. MIETNER, J. BOSSE, and B. THALER. *Hardware In The Loop Simulator Für Rendezvous Und Docking Manöver*. DGLR Jahrestagung (PDF available online). Aachen, Germany, 2009 (cited on page 40).

[19] T. BOGE and E. SCHREUTELKAMP. "A New Command and Control Environment for Rendezvous and Docking Simulations at the EPOS-Facility." In: *7th International Workshop on Simulation for European Space Programmes* (2002) (cited on page 37).

[20] D. C. BROWN. "Decentering Distortion of Lenses." In: *Photometric Engineering* 32.3 (1966), pages 444–462 (cited on page 48).

[21] R. BRUNELLI. *Template Matching Techniques in Computer Vision: Theory and Practice*. John Wiley & Sons, 2009. ISBN: 978-0-470-51706-2 (cited on page 116).

[22] J. F. CANNY. "A Computational Approach To Edge Detection." In: *IEEE Transactions on Pattern Analysis and Machine Intelligence* 8 (6 1986), pages 679–698 (cited on page 66).

[23] J. CHEN, Y. ZHOU, and C. LI. "Formation flying orbit design for the distributed synthetic aperture radar satellite." In: *Science in China Series F: Information Sciences* 47 (6 2004), pages 741–751. ISSN: 1009-2757. DOI: `10.1360/03yf0210` (cited on page 22).

[24] X. CLERC, D. BERTHELIER, M. CHAIZE, H. CLERC, E. MEUNIER, M. ZINK, and S. STRANDMOE. "Qualification of the Automated Transfer Vehicle (ATV) Flight Control." In: *Proceedings of the 7th International ESA Conference on Guidance, Navigation & Control Systems*. 2008. DOI: 10.2514/6.2↩ 006-7266 (cited on page 41).

[25] A. CROPP. *Pose Estimation and Relative Orbit Determination of a Nearby Target Microsatellite Using Passive Imagery*. University of Surrey, 2001 (cited on page 116).

[26] S. CZERNIK. "Design of the Thermal Control System for Compass-1." Diploma thesis. University of Applied Sciences, Aachen, Germany, 2004 (cited on page 199).

[27] W. DAHMEN and A. REUSKEN. *Numerik Für Ingenieure und Naturwissenschaftler*. Springer-Lehrbuch. Springer London, Limited, 2008. ISBN: 9783540764939 (cited on pages 87, 123).

[28] S. D'AMICO, J.-S. ARDAENS, G. GAIAS, B. SCHLEPP, H. BENNINGHOFF, T. TZSCHICHHOLZ, T. KARLSSON, and J. L. JØRGENSEN. "Flight Demonstration of Non-Cooperative Rendezvous using Optical Navigation." In: *23th International Symposium on Space Flight Dynamics*. Pasadena, CA, USA, 2012 (cited on pages 22, 205).

[29] J. DIEBEL and S. THRUN. "An Application of Markov Random Fields to Range Sensing." In: *Proceedings of Conference on Neural Information Processing Systems (NIPS)*. Cambridge, MA: MIT Press, 2005 (cited on page 131).

[30] F. DORNAIKA and B. RADUCANU. "Single snapshot-based 3D head pose initialization for tracking in a HRI scenario." In: *IEEE Computer Society Conference on Computer Vision and Pattern Recognition Workshops (CVPRW)*. June 2010, pages 32–39. DOI: 10.1109/CVPRW.2010.5543786 (cited on page 116).

[31] D. DROESCHEL, D. HOLZ, and S. BEHNKE. "Multi-frequency Phase Unwrapping for Time-of-Flight cameras." In: *International Conference on Intelligent Robots and Systems*. 2010, pages 1463–1469. DOI: 10.1109↩ /IROS.2010.5649488 (cited on page 106).

[32] D. DROESCHEL, D. HOLZ, and S. BEHNKE. "Probabilistic Phase Unwrapping for Time-of-Flight Cameras." In: *ISR/ROBOTIK 2010, Proceedings for the joint conference of ISR 2010 (41st Internationel Symposium on Robotics) und ROBOTIK 2010 (6th German Conference on Robotics)*. VDE Verlag, 2010. ISBN: 978-3-8007-3273-9 (cited on page 106).

[33] T. DRUMMOND and R. CIPOLLA. "Real-Time Tracking of Highly Articulated Structures in the Presence of Noisy Measurements." In: *Proceedings of the International Conference on Computer Vision*. 2001, pages 315–320 (cited on page 137).

[34] T. DRUMMOND and R. CIPOLLA. "Real-Time Visual Tracking of Complex Structures." In: *IEEE Transactions on Pattern Analysis and Machine Intelligence* 24.7 (2002), pages 932–946. ISSN: 0162-8828. DOI: 10.11↩ 09/TPAMI.2002.1017620 (cited on pages 133, 137).

[35] R. O. DUDA and P. E. HART. "Use of the Hough transformation to detect lines and curves in pictures." In: *Communications of the ACM* 15.1 (1972), pages 11–15. ISSN: 0001-0782. DOI: 10.1145/361237.361242 (cited on page 133).

[36] D. ECK, M. BAUMANN, L. LEMMER, and K. SCHILLING. "UKF Sensor Data Fusion for Localisation of a Mobile Robot." In: *Proceedings of the 41st International Symposium on Robotics (ISR)*. Munich, Germany, 2010, pages 362–369 (cited on pages 129, 148).

[37] C. ENGLISH, S. ZHU, C. SMITH, S. RUEL, and I. CHRISTIE. "TriDAR: A hybrid sensor for exploiting the complementary nature of triangulation and LiDAR technologies." In: *ISAIRAS 2005 Conference, ESA SP-603*. Munich, Germany, Sept. 2005 (cited on pages 22, 23, 194).

[38] M. A. FISCHLER and R. C. BOLLES. "Random sample consensus: a paradigm for model fitting with applications to image analysis and automated cartography." In: *Communications of the ACM* 24.6 (June 1981), pages 381–395. ISSN: 0001-0782. DOI: 10.1145/358669.358692 (cited on pages 133, 140).

[39] S. FOIX, G. ALENYA, and C. TORRAS. "Lock-in Time-of-Flight (ToF) Cameras: A Survey." In: *IEEE SENSORS JOURNAL* 11.9 (Sept. 2011), pages 1917–1926. DOI: 10.1109/JSEN.2010.2101060 (cited on pages 49, 100).

[40] F. FRANK. "Orientation mapping." In: *Metallurgical Transactions A- Physical Metallurgy and Materials Science* 19 (3 1988), pages 403–408. ISSN: 0360-2133. DOI: 10.1007/BF02649253 (cited on page 62).

[41] B. J. FREY, R. KOETTER, and N. PETROVIC. "Very loopy belief propagation for unwrapping phase images." In: *Advances in Neural Information Processing Systems 14*. 2001, pages 737–743 (cited on page 106).

[42] J. FREY. *Bildsensorik mit Tiefgang: Stand der PMD-Technik, Roadmap & Perspektiven*. PMDVisionDay 2008, Munich. 2008. http://spect↩ronet.de/portals/visqua/story_docs/vortraege_2008/081118↩_pmd_visionday/081118_01_frey_pmd_technologies.pdf (cited on page 52).

[43] G. FRIEDLAND, K. JANTZ, T. LENZ, and R. ROJAS. *Extending the SIOX Algorithm: Alternative Clustering Methods, Sub-pixel Accurate Object Extraction from Still Images, and Generic Video Segmentation*. Technical report B 06-06. Freie Universität Berlin, 2006 (cited on page 132).

[44] G. FRIEDLAND, K. JANTZ, and R. ROJAS. "SIOX: Simple Interactive Object Extraction in Still Images." In: *International Symposium on Multimedia* (2005), pages 253–260. DOI: 10.1109/ISM.2005.106 (cited on page 132).

[45] S. FUCHS. "Multipath Interference Compensation in ToF-Camera Images." In: *Proceedings of the International Conference on Pattern Recognition 2010, Istanbul, Turkey*. Aug. 2010 (cited on page 94).

[46] S. FUCHS and G. HIRZINGER. "Extrinsic and depth calibration of ToF-cameras." In: *Proceedings of the 2008 IEEE Computer Society Conference on Computer Vision and Pattern Recognition (CVPR 2008), 24-26 June 2008, Anchorage, Alaska, USA*. IEEE Computer Society, 2008. DOI: 10.1109↩/CVPR.2008.4587828 (cited on page 93).

[47] D. GHIGLIA and M. PRITT. *Two-dimensional phase unwrapping: theory, algorithms, and software.* Wiley-Interscience publication. Wiley, 1998. ISBN: 9780471249351 (cited on page 106).

[48] P. GIL, J. POMARES, and F. TORRES. "Analysis and Adaptation of Integration Time in PMD Camera for Visual Servoing." In: *Proceedings of the International Conference on Pattern Recognition.* IEEE, 2010, pages 311–315. DOI: 10.1109/ICPR.2010.85 (cited on pages 163, 171).

[49] V. GOIFFON, P. MAGNAN, F. BERNARD, G. ROLLAND, O. SAINT-PÉ, N. HUGER, and F. CORBIÈRE. "Ionizing radiation effects on CMOS imagers manufactured in deep submicron process." In: *Proceedings of SPIE. Sensors, Cameras, and Systems for Industrial/Scientific Applications IX.* Volume 6816. 2008. DOI: 10.1117/12.767484 (cited on page 200).

[50] A. GOTTSCHEBER and S. DECH. "Automatic detection of space debris with a meade telescope." In: *Proceedings of the 61st International Astronautical Congress.* 2010 (cited on pages 19, 20).

[51] D. GREIG, B. PORTEOUS, and A. SEHEULT. "Exact Maximum A Posteriori Estimation for Binary Images." In: *Journal of the Royal Statistical Society Series B* 51 (1989), pages 271–279 (cited on page 131).

[52] S. Á. GUÐMUNDSSON, H. AANÆS, and R. LARSEN. "Fusion of stereo vision and Time-Of-Flight imaging for improved 3D estimation." In: *International Journal of Intelligent Systems Technologies and Applications* 5.3/4 (2008), pages 425–433. ISSN: 1740-8865. DOI: 10.1504/IJISTA.2008.021305 (cited on page 131).

[53] G. GUGLIERI, F. MAROGLIO, P. PELLEGRINO, and L. TORRE. "A Ground Facility to Test GNC Algorithms and Sensors for Autonomous Rendezvous and Docking." In: *Proceedings of the 1st IAA conference on Dynamics and Control of Space Systems.* 2012 (cited on page 41).

[54] U. HAHNE and M. ALEXA. "Combining Time-Of-Flight depth and stereo images without accurate extrinsic calibration." In: *International Journal of Intelligent Systems Technologies and Applications* 5.3/4 (2008), pages 325–333. ISSN: 1740-8865. DOI: 10.1504/IJISTA.2008.021295 (cited on page 131).

[55] C. HARRIS. "Tracking with rigid models." In: *Active vision*. Cambridge, MA, USA: MIT Press, 1993, pages 59–73. ISBN: 0-262-02351-2 (cited on page 137).

[56] F. HASOUNEH, S. KNEDLIK, V. PETERS, and O. LOFFELD. "PMD-based sensor node position monitoring." In: *PLANS 2006 (IEEE/ION Positioning Location and Navigation Symposium)*. Coronado (San Diego), USA, 2006, pages 569–573 (cited on page 192).

[57] E. HECHT. *Optics*. Addison-Wesley, 2002. ISBN: 9780805385663 (cited on page 42).

[58] H. G. HEINOL. "Untersuchung und Entwicklung von modulationslaufzeit-basierten 3D-Sichtsystemen." Dissertation. University of Siegen, 2001 (cited on page 51).

[59] U. HILLENBRAND and R. LAMPARIELLO. "Motion and Parameter Estimation of a Free-Floating Space Object from Range Data for Motion Prediction." In: *Proceedings of the 8th International Symposium on Artificial Intelligence, Robotics and Automation in Space (i-SAIRAS)*. 2005 (cited on page 133).

[60] B. HUHLE, P. JENKE, and W. STRASSER. "On-the-Fly Scene Acquisition with a Handy Multisensor-System." In: *Dynamic 3D Imaging Workshop in Conjunction with DAGM 2007*. Heidelberg, Germany, Sept. 2007 (cited on pages 130, 131, 148).

[61] B. JÄHNE. *Digitale Bildverarbeitung und Bildgewinnung (7. Auflage)*. Springer Vieweg, Berlin Heidelberg, 2012. ISBN: 978-3-642-04951-4. DOI: 10.1007/978-3-642-04952-1 (cited on page 137).

[62] P. JASIOBEDZKI, M. ABRAHAM, P. NEWHOOK, and J. TALBOT. "Model based pose estimation for autonomous operations in space." In: *Proceedings of the International Conference on Information Intelligence and Systems*. 1999, pages 211–215. DOI: 10.1109/ICIIS.1999.810263 (cited on page 195).

[63] P. JASIOBEDZKI, S. SE, T. PAN, M. UMASUTHAN, and M. GREENSPAN. "Autonomous satellite rendezvous and docking using LIDAR and model based vision." In: *Proceedings of SPIE 5798, Spaceborne Sensors II*. 2005, pages 54–65. DOI: 10.1117/12.604011 (cited on pages 23, 41, 116, 194, 196).

[64] C. JOOCHIM and H. ROTH. "Development of a 3D mapping using 2D/3D sensors for mobile robot locomotion." In: *IEEE International Conference on Technologies for Practical Robot Applications*. 2008, pages 100–105. DOI: 10.1109/TEPRA.2008.4686681 (cited on page 130).

[65] C. JOOCHIM and H. ROTH. "Mobile Robot Exploration Based on Three Dimension Cameras Acquisition." In: *2nd IFAC Symposium on Telematics Applications*. 2010, pages 116–121. DOI: 10.3182/20101005-4-RO-2↩ 018.00037 (cited on pages 130, 131).

[66] C. KAISER, F. SJÖBERG, J. M. DELCURA, and B. EILERTSEN. "SMART-OLEV- An orbital life extension vehicle for servicing commercial spacecrafts in GEO." In: *Acta Astronautica* 63.1-4 (2008). Touching Humanity- Space for Improving Quality of Life. Selected Proceedings of the 58th International Astronautical Federation Congress, Hyderabad, India, 24-28 September 2007, pages 400–410. ISSN: 0094-5765. DOI: 10.1016/j.actaastro.2007↩ .12.053 (cited on pages 21, 163).

[67] R. KASPER and S. SCHMIDT. "Sensor-data-fusion for an autonomous vehicle using a Kalman-filter." In: *6th International Symposium on Intelligent Systems and Informatics*. 2008, pages 1–5. DOI: 10.1109/SISY.2008.↩ 4664905 (cited on pages 129, 148).

[68] I. KAWANO, H. UENO, Y. ISHIJIMA, T. ADACHI, and T. IIJIMA. "Study on Laser Rangefinder for Non-cooperative Rendezvous." In: *Proceedings of the International Symposium on Artificial Intelligence, Robotics and Automation in Space*. Japan, 2003 (cited on page 116).

[69] J. KELSEY, J. BYRNE, M. COSGROVE, S. SEEREERAM, and R. MEHRA. "Vision-based relative pose estimation for autonomous rendezvous and docking." In: *Proceedings of the IEEE Aerospace Conference*. 2006. DOI: 10.1109/A↩ ERO.2006.1655916 (cited on pages 133, 195, 196).

[70] P. KHONGSAB. "Signal processing and performance evaluation of a PMD camera for space docking." Master's thesis. Germany: University of Würzburg, Faculty of Computer Science, Institute of Robotics and Telematics, 2009 (cited on page 192).

[71] W. KIM, W. YIBING, I. OVSIANNIKOV, S. LEE, Y. PARK, C. CHUNG, and E. FOSSUM. "A 1.5Mpixel RGBZ CMOS image sensor for simultaneous color and range image capture." In: *Solid-State Circuits Conference Digest of Technical Papers (ISSCC)*. 2012, pages 392–394. DOI: 10.1109/ISS↩CC.2012.6177061 (cited on page 207).

[72] A. KOLB, E. BARTH, R. KOCH, and R. LARSEN. "Time-of-Flight Sensors in Computer Graphics." In: *Proceedings of Eurographics 2009- State of the Art Reports*. The Eurographics Association, 2009, pages 119–134 (cited on page 69).

[73] P. KOVESI. "A Dimensionless Measure of Edge Significance." In: *Conference on Digital Image Computing: Techniques and Applications*. 1991, pages 281–288 (cited on page 66).

[74] H. KRAFT, J. FREY, T. MOELLER, M. ALBRECHT, M. GROTHOF, B. SCHINK, H. HESS, and B. BUXBAUM. *3D-Camera of High 3D-Frame Rate, Depth-Resolution and Background Light Elimination Based on Improved PMD (Photonic Mixer Device)-Technologies*. Technical report. Universität Siegen, 2004 (cited on page 52).

[75] J. KREBS, P. PISSAVIN, and D. VILAIRE. "SED 16 Autonomous Star Tracker." In: *Proceedings of the 4th International ESA Conference on Spacecraft Guidance, Navigation and Control Systems*. 2000, pages 569–574 (cited on page 44).

[76] K.-D. KUHNERT and M. STOMMEL. "Fusion of Stereo-Camera and PMD-Camera Data for Real-Time Suited Precise 3D Environment Reconstruction." In: *IEEE/RSJ International Conference on Intelligent Robots and Systems*. Oct. 2006, pages 4780–4785. DOI: 10.1109/IROS.2006.282↩349 (cited on page 131).

[77] G. LANDIS, S. BAILEY, and R. TISCHLER. "Causes of Power-Related Satellite Failures." In: *Conference Record of the 2006 IEEE 4th World Conference on Photovoltaic Energy Conversion*. Volume 2. May 2006, pages 1943–1945. DOI: 10.1109/WCPEC.2006.279878 (cited on page 21).

[78] R. LANGE. "3D time-of-flight distance measurement with custom solid-state image sensors in CMOS/CCD-technology." Disseration. University of Siegen, 2000 (cited on pages 56, 57, 97).

[79] D. LAU. *Algebra und Diskrete Mathematik 1*. Springer, 2004. ISBN: 3-540-20397-4 (cited on page 123).

[80] J. LEE, C. CARRINGTON, S. SPENCER, T. BRYAN, R. T. HOWARD, and J. JOHNSON. "Next Generation Advanced Video Guidance Sensor: Low Risk Rendezvous and Docking Sensor." In: *MSFC-2080- AIAA Space 2008* (2008) (cited on pages 24, 194).

[81] S. LEE. "Depth camera image processing and applications." In: *19th IEEE International Conference on Image Processing*. 2012, pages 545–548. DOI: 10.1109/ICIP.2012.6466917 (cited on pages 49, 52).

[82] V. LEPETIT and P. FUA. "Monocular Model-Based 3D Tracking of Rigid Objects: A Survey." In: *Foundations and Trends in Computer Graphics and Vision*. 2005, pages 1–89 (cited on page 193).

[83] S. LEWIS, S. SÖCHTING, E. ROBERTS, and I. SHERRINGTON. "Observations on the influence of launch vibration on bearing torque and lubricant performance." In: *9th European Space Mechanisms and Tribology Symposium*. Edited by R. HARRIS. Volume 480. ESA Special Publication. Sept. 2001, pages 271–278 (cited on page 201).

[84] S. Z. LI. *Markov random field modeling in image analysis*. Springer-Verlag New York, Inc., Secaucus, NJ, USA, 2001 (cited on page 131).

[85] Z.-P. LIANG and P. C. LAUTERBUR. *Principles of Magnetic Resonance Imaging: A Signal Processing Perspective*. Wiley-IEEE Press, 1999. ISBN: 0780347234 (cited on page 106).

[86] M. LINDNER and A. KOLB. "Calibration of the intensity-related distance error of the PMD TOF-Camera." In: *SPIE: Intelligent Robots and Computer Vision XXV*. Volume 6764. 2007. DOI: 10.1117/12.752808 (cited on pages 93, 94, 100).

[87] M. LINDNER, A. KOLB, and K. HARTMANN. "Data Fusion of PMD-Based Distance Information and High Resolution RGB-Images." In: *International Symposium on Signals, Circuits and Systems*. Volume 1. July 2007. DOI: 10.1109/ISSCS.2007.4292666 (cited on pages 130, 131).

[88] M. LINDNER and A. KOLB. "Lateral and Depth Calibration of PMD-Distance Sensors." In: *ISVC (2)*. Edited by G. BEBIS et al. Volume 4292. Lecture Notes in Computer Science. Springer, 2006, pages 524–533. ISBN: 3-540-48628-3. DOI: 10.1007/11919629_53 (cited on page 93).

[89] M. LINDNER, M. LAMBERS, and A. KOLB. "Sub-pixel data fusion and edge-enhanced distance refinement for 2D/3D images." In: *International Journal of Intelligent Systems Technologies and Applications* (5 2008), pages 344–354. DOI: 10.1504/IJISTA.2008.021297 (cited on pages 130, 132).

[90] M. LINDNER, I. SCHILLER, A. KOLB, and R. KOCH. "Time-of-Flight sensor calibration for accurate range sensing." In: *Computer Vision and Image Understanding* 114.12 (2010). Special issue on Time-of-Flight Camera Based Computer Vision, pages 1318–1328. ISSN: 1077-3142. DOI: 10.1016/↩ j.cviu.2009.11.002 (cited on pages 93, 97).

[91] A. M. LONG, M. G. RICHARDS, and D. E. HASTINGS. "On-Orbit Servicing: A New Value Proposition for Satellite Design and Operation." In: *Journal of Spacecraft and Rockets* 44 (4 2007), pages 964–976. DOI: 10.2514/1.2↩ 7117 (cited on pages 20, 21).

[92] M. LOURAKIS. *levmar: Levenberg-Marquardt nonlinear least squares algorithms in C/C++*. web page. 2004. http://www.ics.forth.gr/~lour↩ akis/levmar/ (cited on pages 146, 148).

[93] D. G. LOWE. "Distinctive Image Features from Scale-Invariant Keypoints." In: *International Journal of Computer Vision* 60 (2 Nov. 2004), pages 91–110. ISSN: 0920-5691. DOI: 10.1023/B:VISI.0000029664.99615.94 (cited on pages 133, 206).

[94] T. LUHMANN. *Nahbereichsphotogrammetrie*. Wichmann Verlag, 2003 (cited on pages 47, 130, 148).

[95] C. MIRAVET, L. PASCUAL, E. KROUCH, and J. M. DEL CURA. "An image-based sensor system for autonomous rendez-vous with uncooperative satellites." In: *Proceedings of the 7th International ESA Conference on Guidance, Navigation & Control Systems*. 2008 (cited on page 132).

[96] T. MÖLLER, H. KRAFT, J. FREY, M. ALBRECHT, and R. LANGE. "Robust 3D Measurement with PMD Sensors." In: *Proceedings of the 1st Range Imaging Research Day at ETH*. 2005 (cited on page 50).

[97] J. J. MORÉ. "The Levenberg-Marquardt algorithm: Implementation and theory." In: *Numerical Analysis, Lecture Notes Math. 630* (1978). Edited by G. A. WATSON, pages 105–116 (cited on page 146).

[98] C. NETRAMAI, O. MELNYCHUK, J. CHANIN, and H. ROTH. "Combining PMD and Stereo camera for Motion Estimation of a Mobile Robot." In: *Proceedings of the 17th IFAC World Congress*. Volume 17. Seoul, Korea, 2008 (cited on page 131).

[99] R. OLFATI-SABER. "Distributed Kalman filtering and sensor fusion in sensor networks." In: *Network Embedded Sensing and Control, volume LNCIS 331*. Springer-Verlag, 2006, pages 157–167 (cited on pages 129, 148).

[100] N. W. OUMER and G. PANIN. "Tracking And Pose Estimation Of Non-Cooperative Satellite For On-Orbit Servicing." In: *Proceedings of the International Symposium on Artificial Intelligence, Robotics and Automation in Space*. 2012 (cited on pages 195, 196).

[101] N. OUMER and G. PANIN. "3D point tracking and pose estimation of a space object using stereo images." In: *Proceedings of the 21st International Conference on Pattern Recognition*. 2012, pages 796–800 (cited on page 22).

[102] T. PATTINSON. "Quantification and Description of Distance Measurement Errors of a Time-of-Flight Camera." Master's thesis. University of Stuttgart, 2010 (cited on page 97).

[103] S. PERSSON, P. BODIN, E. GILL, J. HARR, and J. JÖRGENSEN. "PRISMA- An autonomous formation flying mission." In: *Proceedings of the ESA Small Satellite Systems and Services Symposium (4S)*. Sardinia, Italy, 2006 (cited on pages 22, 163).

[104] N. PHILIP and M. ANANTHASAYANAM. "Relative position and attitude estimation and control schemes for the final phase of an autonomous docking mission of spacecraft." In: *Acta Astronautica* 52.7 (2003), pages 511–522. ISSN: 0094-5765. DOI: 10.1016/S0094-5765(02)00125-X (cited on pages 193, 196).

[105] D. PIATTI and F. RINAUDO. "SR-4000 and CamCube3.0 Time of Flight (ToF) Cameras: Tests and Comparison." In: *Remote Sensing* 4.4 (2012), pages 1069–1089. ISSN: 2072-4292. DOI: 10.3390/rs4041069 (cited on page 49).

[106] T. PRASAD, K. HARTMANN, W. WOLFGANG, S. GHOBADI, and A. SLUITER. "First steps in enhancing 3D vision technique using 2D/3D sensors." In: *11. Computer Vision Winter Workshop 2006*. Edited by V. CHUM O.Franc. Telc, Czech Republic: Czech Society for Cybernetics and Informatics, 2006, pages 82–86. ISBN: 80-239-6530-1 (cited on page 131).

[107] F. QURESHI and D. TERZOPOULOS. "Intelligent perception and control for space robotics: Autonomous Satellite Rendezvous and Docking." In: *Machine Vision and Applications* 19.3 (Feb. 2008), pages 141–161. ISSN: 0932-8092. DOI: 10.1007/s00138-007-0085-z (cited on page 23).

[108] J. RADMER, P. FUSTE, H. SCHMIDT, and J. KRUGER. "Incident light related distance error study and calibration of the PMD-range imaging camera." In: *Conference on Computer Vision and Pattern Recognition Workshops*. June 2008. DOI: 10.1109/CVPRW.2008.4563168 (cited on page 93).

[109] H. RAPP. "Experimental and Theoretical Investigation of Correlating TOF-Camera Systems." Diploma thesis. University of Heidelberg, 2007 (cited on pages 51, 92).

[110] L. REGOLI, K. RAVANDOOR, M. SCHMIDT, and K. SCHILLING. "Advanced Techniques for Spacecraft Motion Estimation Using PMD Sensors." In: *Proceedings of the 1st Conference on Embedded Systems, Computational Intelligence and Telematics in Control*. 2012, pages 320–325. DOI: 10.↩ 3182/20120403-3-DE-3010.00079 (cited on page 132).

[111] L. REGOLI, K. RAVANDOOR, M. SCHMIDT, and K. SCHILLING. "On-line robust pose estimation for Rendezvous and Docking in space using Photonic Mixer Devices." In: *Proceedings of the 63th International Astronautical Congress.* 2012 (cited on pages 133, 189).

[112] L. REGOLI. "Advanced attitude determination using Photonic Mixer Devices." Master's thesis. Germany: University of Würzburg, Faculty of Computer Science, Institute of Robotics and Telematics, 2011 (cited on pages 132, 192).

[113] F. REMS. "Integration Of The Formation Flying Testbed With The European Proximity Operations Simulator." In: *61. Deutscher Luft- und Raumfahrtkongress* (2012) (cited on page 156).

[114] R. REULKE. *Combination of distance data with high resolution images.* Technical report. Humboldt-Universität Berlin, 2006 (cited on page 130).

[115] A. RICHARDS and J. P. HOW. "Performance Evaluation Of Rendezvous Using Model Predictive Control." In: *AIAA Guidance, Navigation, and Control Conference (GNC).* Austin, TX, 2003 (cited on pages 75, 93).

[116] C. RICHARDT, C. STOLL, N. A. DODGSON, H.-P. SEIDEL, and C. THEOBALT. "Coherent Spatiotemporal Filtering, Upsampling and Rendering of RGBZ Videos." In: *Proceedings of Eurographics Computer Graphics Forum.* Volume 31. 2. Cagliari, Italy, May 2012. DOI: 10.1111/j.1467-8659.20↩ 12.03003.x (cited on page 130).

[117] G. G. RIGATOS. "Extended Kalman and Particle Filtering for sensor fusion in motion control of mobile robots." In: *Journal of Mathematics and Computers in Simulation* 81.3 (2010), pages 590–607. ISSN: 0378-4754. DOI: 10.1016/j.matcom.2010.05.003 (cited on page 129).

[118] F. D. ROE, R. T. HOWARD, and L. MURPHY. "Automated rendezvous and capture system development and simulation for NASA." In: *SPIE 5420, Modeling, Simulation, and Calibration of Space-based Systems* (2004), pages 118–125. DOI: 10.1117/12.542529 (cited on page 41).

[119] T. RUPP, T. BOGE, R. KIEHLING, and F. SELLMAIER. "Flight Dynamics Challenges Of The German On-Orbit Servicing Mission DEOS." In:

21st International Symposium on Space Flight Dynamics (2009) (cited on page 40).

[120] S. RUSINKIEWICZ and M. LEVOY. "Efficient Variants of the ICP Algorithm." In: *3rd International Conference on 3D Digital Imaging and Modeling (3DIM 2001), 28 May - 1 June 2001, Quebec City, Canada.* IEEE Computer Society, 2001, pages 145–152. ISBN: 0-7695-0984-3. DOI: 10.1109/IM.2001.924423 (cited on page 194).

[121] A. SABOV and J. KRÜGER. "Identification and correction of flying pixels in range camera data." In: *Proceedings of the 24th Spring Conference on Computer Graphics.* SCCG '08. Budmerice, Slovakia: ACM, 2010, pages 135–142. ISBN: 978-1-60558-957-2. DOI: 10.1145/1921264.1921293 (cited on page 94).

[122] J. H. SALEH, D. E. HASTINGS, E. S. LAMASSOURE, and D. J. NEWMAN. "Flexibility and the Value of On-Orbit Servicing: New Customer-Centric Perspective." In: *Journal of Spacecraft and Rockets* 40 (2 2003), pages 279–291. DOI: 10.2514/2.3944 (cited on page 21).

[123] N. SANTRAC, G. FRIEDLAND, and R. ROJAS. *High Resolution Segmentation with a Time-of-Flight 3D-Camera using the Example of a Lecture Scene.* Technical report. Freie Universität Berlin, 2006 (cited on page 132).

[124] S. SAVARESE and L. FEI-FEI. *3D generic object categorization, localization and pose estimation.* Technical report. University of Illinois, Princeton University, 2007 (cited on page 132).

[125] I. SCHILLER, C. BEDER, and R. KOCH. "Calibration of a PMD camera using a planar calibration object together with a multi-camera setup." In: *The International Archives of the Photogrammetry, Remote Sensing and Spatial Information Sciences.* Volume XXXVII. Part B3a. XXI. ISPRS Congress. Beijing, China, 2008, pages 297–302 (cited on page 93).

[126] M. SCHMIDT and K. SCHILLING. "Formation Flying Techniques for Pico-Satellites." In: *6th International Workshop on Satellite Constellation and Formation Flying.* 2010 (cited on page 22).

[127] M. SCHMIDT. "Analysis, Modeling and Dynamic Optimization of 3D Time-of-Flight Imaging Systems." Dissertation. IWR, Fakultät für Physik und Astronomie, Univ. Heidelberg, 2011 (cited on pages 58, 93, 199).

[128] SCIENTIFIC AND TECHNICAL SUBCOMMITTEE OF THE UNITED NATIONS COMMITTEE ON THE PEACEFUL USES OF OUTER SPACE. *Technical Report on Space debris*. United Nations, 1999. ISBN: 92-1-100813-1 (cited on page 20).

[129] M. SEGAL, C. KOROBKIN, R. VAN WIDENFELT, J. FORAN, and P. HAEBERLI. "Fast shadows and lighting effects using texture mapping." In: *SIGGRAPH Computer Graphics* 26.2 (July 1992), pages 249–252. ISSN: 0097-8930. DOI: 10.1145/142920.134071 (cited on page 131).

[130] F. SELLMAIER, T. BOGE, J. SPURMANN, S. GULLY, T. RUPP, and F. HUBER. "On-Orbit Servicing Missions: Challenges and Solutions for Spacecraft Operations." In: *Proceedings of the SpaceOps 2010 Conference* (2010) (cited on page 167).

[131] F. SELLMAIER, J. SPURMANN, and T. BOGE. "On-Orbit Servicing Missions At DLR/ GSOC." In: *Proceedings of the 61st International Astronautical Congress*. 2006 (cited on page 21).

[132] A. SHAHROKNI, T. DRUMMOND, and P. FUA. "Texture Boundary Detection for Real-Time Tracking." In: *European Conference on Computer Vision*. 2004, pages 566–577. DOI: 10.1007/978-3-540-24671-8_45 (cited on pages 66, 137).

[133] G. SHAKHNAROVICH, P. VIOLA, and T. DARRELL. "Fast Pose Estimation with Parameter-Sensitive Hashing." In: *Proceedings of the Ninth IEEE International Conference on Computer Vision - Volume 2*. ICCV '03. Washington, DC, USA: IEEE Computer Society, 2003, pages 750–. ISBN: 0-7695-1950-4 (cited on page 116).

[134] I. SOBEL and G. FELDMAN. "A 3x3 Isotropic Gradient Operator for Image Processing." In: *Pattern Classification and Scene Analysis* (1973), pages 271–273 (cited on page 66).

[135] M. STRENGERT, M. KRAUS, and T. ERTL. "Pyramid Methods in GPU-Based Image Processing." In: *Workshop on Vision, Modelling, and Visualization VMV '06*. 2006, pages 169–176 (cited on page 132).

[136] C. TAMM. "Stored energy electrical connectors." In: *Proceedings of the 9th International IEEE Conference on Transmission and Distribution Construction, Operation and Live-Line Maintenance*. 2000, pages 129–134. DOI: 10.1109/TDCLLM.2000.882810 (cited on page 201).

[137] J.-P. TAREL, H. CIVI, and D. B. COOPER. "Pose Estimation of Free-Form 3D Objects without Point Matching using algebraic surface Models." In: *Proceedings of IEEE Workshop Model Based 3D Image Analysis*. Mumbai, India, 1998, pages 13–21 (cited on page 116).

[138] F. TERUI. "Model based visual relative motion estimation and control of a spacecraft utilizing computer graphics." In: *21st International Symposium on Space Flight dynamics*. Toulouse, France, 2009 (cited on pages 41, 89, 189, 194).

[139] P. THOMPSON, D. E. WAHL, P. H. EICHEL, D. C. GHIGLIA, and C. V. JAKOWATZ. *Spotlight-Mode Synthetic Aperture Radar: A Signal Processing Approach*. Norwell, MA, USA: Kluwer Academic Publishers, 1996. ISBN: 0792396774 (cited on page 106).

[140] G. THUILLIER, M. HERSÉ, D. LABS, T. FOUJOLS, W. PEETERMANS, D. GILLOTAY, P. SIMON, and H. MANDEL. "The Solar Spectral Irradiance from 200 to 2400 nm as Measured by the SOLSPEC Spectrometer from the Atlas and Eureca Missions." In: *Solar Physics* 214 (May 2003), pages 1–22. DOI: 10.1023/A:1024048429145 (cited on page 153).

[141] T. TZSCHICHHOLZ, H. BENNINGHOFF, and T. BOGE. "A flexible image processing framework for vision-based navigation using monocular imaging sensors." In: *Proceedings of the 8th International ESA Conference on Guidance, Navigation & Control Systems*. Carlsbad, Czech Republic, 2011 (cited on pages 43, 74, 77, 193, 205).

[142] T. TZSCHICHHOLZ and T. BOGE. "GNC systems development in conjunction with a RvD hardware-in-the-loop simulator." In: *Proceedings of the 4th International Conference on Astrodynamics Tools & Techniques*. Madrid, Spain, 2010 (cited on pages 43, 184).

[143] T. TZSCHICHHOLZ, L. MA, and K. SCHILLING. "Model-based spacecraft pose estimation and motion prediction using a photonic mixer device camera." In: *Acta Astronautica* 68.7-8 (2011), pages 1156–1167. ISSN: 0094-5765. DOI: 10.1016/j.actaastro.2010.10.003 (cited on pages 116, 133, 163, 186, 189, 192, 204, 206).

[144] T. TZSCHICHHOLZ and K. SCHILLING. "Range extension of the PMD sensor with regard to applications in space." In: *Proceedings of the 19th IFAC Symposium on Automatic Control in Aerospace*. 2013, pages 324–329. DOI: 10.3182/20130902-5-DE-2040.00004 (cited on page 114).

[145] *VDM- Videometer*. Product brief. 2004. http://www.sodern.com/sit↵es/docs_wsw/RUB_54/VDM.pdf (cited on page 44).

[146] S. VERES and N. LINCOLN. "Vision Assisted Satellite Formation Control." In: *Proceedings of the 45th IEEE Conference on Decision and Control*. Dec. 2006, pages 5712–5717. DOI: 10.1109/CDC.2006.377489 (cited on page 22).

[147] J. R. WERTZ and R. BELL. "Autonomous Rendezvous and Docking Technologies- Status and Prospects." In: *Proceedings of the SPIE AeroSense Symposium*. Volume 5088-3. 2003 (cited on page 21).

[148] J. WERTZ. *Spacecraft Attitude Determination and Control*. Astrophysics and Space Science Library. Reidel, 1978. ISBN: 9789027709592 (cited on page 30).

[149] M. WIEDEMANN, M. SAUER, F. DRIEWER, and K. SCHILLING. "Analysis and characterization of the PMD camera for application in mobile robotics." In: *The 17th IFAC World Congress, Seoul, Korea*. 2008, pages 13689–13694 (cited on pages 56, 93, 163, 171).

[150] T. WIMMER and T. BOGE. "EPOS: A Hardware-in-the-loop Robotic Simulation Assembly for Testing Automated Rendezvous and Docking GNC sensor payloads." In: *Proceedings of the 8th International ESA Conference on Guidance, Navigation & Control Systems*. 2011 (cited on page 40).

[151] P. WUNSCH, S. WINKLER, and G. HIRZINGER. "Real-time pose estimation of 3D objects from camera images using neural networks." In: *Proceedings of the IEEE International Conference on Robotics and Automation 1997*.

Volume 4. Apr. 1997, pages 3232–3237. DOI: 10.1109/ROBOT.1997.6↩
06781 (cited on page 116).

[152] X. YIN and N. NOGUCHI. "Motion Detection and Tracking Using the 3D-camera." In: *Proceedings of the 18th IFAC World Congress*. 2011, pages 14139–14144. DOI: 10.3182/20110828-6-IT-1002.01762 (cited on page 132).

[153] M. ZEBENAY, R. LAMPARIELLO, T. BOGE, and D. CHOUKROUN. "A New Contact Dynamics Model Tool For Hardware-In-The-Loop Docking Simulation." In: *Proceedings of the International Symposium on Artificial Intelligence, Robotics and Automation in Space*. 2012 (cited on page 39).

[154] S. ZHANG and X. CAO. "Closed-form solution of monocular vision-based relative pose determination for RVD spacecrafts." In: *Aircraft Engineering and Aerospace Technology: An International Journal* 77.3 (Mar. 2005), pages 192–198. ISSN: 0002-2667. DOI: 10.1108/00022660510597214 (cited on pages 193, 196).

Index

A

Accuracy comparison, 197
Actuators, 19
Algebraic surface model, 116
Algorithm limits, 205
Ambiguity
    Coefficients, 107
    Indicator, 108
    Interval resolution, 106
Amplitude (PMD), 55
    Distance error compensation, 102
Amplitude threshold (coarse acquisition), 120
ARGON experiment, 22
Autocorrelation function, 52
Autonomous Transfer Vehicle, 38, 44
AVGS, 24

B

Barycentric coordinates, 146
Bootstrap point, 120

C

Caltech Camera Calibration Toolbox, 48

CAM, 154
CameraLink, 44
Canny algorithm, 66
CCD camera, 43, 46
    Calibration, 46–49
    Characteristics, 73
    Color, 44
    Interface, 44–45
    Latency, 158
    Mechanical and power, 46
    Optical calibration results, 48
    Optics, 45
    Resolution, 44
    Selection criteria, 44–46
CEV, 24
Coarse acquisition, 117
Command frequency (EPOS), 37
Communication delay, 21
Complex plane, 54
Contact dynamics, 39
Contributions, 24–26
Conventions, 30
Corner threshold, 83

D

Data fusion, 131
Decaying orbit, 21
Deorbiting, 21
DEOS mission, 21, 209
Depth perception, 74
Direct distance ambiguity estimation, 110
  Noise-range trade-off, 112
Direction vectors, 88
Distance (PMD), 55
Distance threshold (coarse acquisition), 120
DLR, 36
Dominant surface (definition), 136

E

Eclipse phase, 192
Edge tracking, 66, 74
EPOS
  ACS, 39
  Architecture, 39
  Contact dynamics, 39
  Dimensions, 36
  FMC, 39
  Introduction, 35–40
  Onboard Measurement System, 39, 165
  Performance summary, 40
Error metric, 149
ESA, 36
EtherCAT, 39
Ethernet, 44
Euler angles, 148

Order, 30, 167
EUROSIM, 37
Extrinsic calibration, 60–65
  Angular refinement, 64
  Results, 65

F

Feature detector
  SIFT, 135
  SURF, 135
Feature-based initialization, 116
Field of view (FOV)
  CCD camera, 46
  PMD camera, 52
Firewire, 44
Fixed pattern noise (PMD), 91, 94–96
Flying pixels, *see* Jump edges
Focal length, 46
Focal ratio, 46
Formation flying, 22–23
Frame rate, 75

G

Geometric feature estimation, 136–138
GigEVision, 44
Gimbal lock, 148
GPS, 22
Graveyard orbit, 21
GSOC, 36

H

Hough transform, 116, 135

I

Image processing methods, 66–69

Image space deviation components, 143, 146
Image space (definition), 43
Initializer, 25
    Hints, 25
Inner edge (definition), 81
Integration time (PMD)
    Distance error, 92
    Distance error compensation, 100–102
Intensity (PMD), 56
    Distance error, 92, 94
International Space Station, 38
Iterative Closest Point Algorithm, 196

J

Jump edges, 93–94

K

Kinect sensor, 199

L

Lambertian reflectance, 94
Laser scanners, 196
Latency
    Satellite communication, 21
    Sensors, 158
Lateral resolution, 74
    CCD, 44
    PMD, 49
Levenberg-Marquardt algorithm, 148, 150
LiDAR, 22, 23
Line (definition), 81

M

Markers, *see* Visual aids, 22, 24
Markov Random Fields, 133
MATLAB, 159
Mockup interface, 40
Model preprocessing, 78
Modulation frequency selection, 111
Moment of area, 85
Monocular vision, 195
Motion artifacts, 92–93
Motivation, 19
Multi-frequency imaging, 106
    Noise considerations, 110
    Specifications, 112
Multipath interference, 92–94

N

Nearest neighbor search, 116
Networked satellites, 22
Neural network, 116
Noise, 75, 110
Noise-range trade-off (DDAE), 112–113
Normal identity margin, 80
Normalized coordinates, 42
Notation, 28–30

O

Object recognition, 133
Object symmetry resolution, 127–128
Occlusion, 116
OLEV mission, 20
On-board autonomy, 21
On-Orbit Servicing, 20–22

OpenGigEVision, 45
Optical distortion, 42, 43, 47, 60, 95
Optical splitter, 133
Orbital Express mission, 24
Orthophoto method, 132
Outer edge (definition), 81

P

Performance comparison, 197
Performance measurements
    Setup, 153
Perspective projection, 42
Phase congruency (edge detection), 66
Phase shift, 30
Phase unwrapping, 106
Pinhole camera model, 42
Pixel grid, 47
Planetary landing, 23
PMD camera, 22, 49–58
    Amplitude, 55
    Characteristics, 73
    Data consistency checking, 58
    Data preprocessing, 91
    Definition, 30
    Distance, 55
    Energy consumption, 22
    Firmware problem, 51
    Fixed pattern noise correction, 94–96
    Intensity, 56
    Latency, 158
    Measurement errors, 91
    Measurement process, 107
    Multiple exposures, 50
    Optical calibration results, 59
    Range ambiguity, 107
    Range extension, 25, 105–113
    SBI, 56, 94
    Selection criteria, 49–50
    Space readiness, 24, 26
    Specifications, 52
    Theory of operation, 51–58
Point cloud, 136, 145
Pose
    Estimation, 21–22, 134, 147
    Representation, 148
    Search space separation, 76, 78
Pose vector (definition), 148
Principal Component Analysis, 86, 116, 122
Principal Point, 47
Principal point, 43
PRISMA satellites, 22, 207
Projective texture map, 133
PSH, 116

Q

Quantization, 69, 141
Quaternion, 148

R

Radiation considerations, 202
RANSAC, 142, 146
Raytracing, 145
Reference frame, 64
Reference measurements, 163
Reference sun spectrum, 156
Reflectors, *see* Visual aids, 24

Rendezvous and Docking, 23–24, 37, 93, 116
    Phases, 21
    Testbed, 24, 26, 35, 41
Rendezvous scenarios, 184
Reprojection error, 49
Rodrigues rotation vector, 62
Rotation matrix, 148
Rover navigation, 23

S

Satellite
    Cluster, 22
    Construction, 22
    Failure, 21
    Fuel supply, 20
    Lifespan, 20
Satellite body frame, 64
SBI (PMD), 50, 56, 94
Scanline, 66
    Parameters, 139
Scanline algorithm, 137, 145
Scanline (definition), 139
Scene acquisition, 132
Scene segmentation, 133
Sensor latency, 158
Sensor mounting, 154
Signal quality indicator, 55
SimCon block, 158
Simulink, 159
Singular Value Decomposition, 87, 123
SIOX algorithm, 134
Skew, 47
Sobel filter, 66

Solar irradiance, 26, 156
Space debris, 19–20
    Detection, 20
    Relative velocities, 20
Spectral response
    CCD, 156
Spectrometer details, 155
State diagram (estimator), 118
Stereo vision, 135, 196
Straightness threshold, 83
Subpixel, 43
Sun simulation, 155, 184
Surface affinity determination, 145
Surface affinity matrix determination, 121
Surface affinity matrix (definition), 136
Surface area threshold, 80
Surface bounding box, 88
Surface constraint margin, 80
Surface extraction, 79
Surface size deviation threshold, 125
Synthetic modulation frequency, 112

T

Target
    Acquisition, 115–128
    High frequency structures, 205
    Size considerations, 205
    Visibility considerations, 205
Template matching, 116
Texture change point, 141
Texture probability, 142
Texture segmentation, 66–69, 141

Thermal considerations, 201
Thermal equilibrium (PMD), 93
Time of flight, 23, 30
Tracker, 26
Tracking, 131
Transition zones, 110
Triangulation, 23
TriDAR, 23

U

Unambiguous interval (definition), 105
Uncooperative target, 19
USB, 44

V

Vibration considerations, 202
Visual aids, 19
VME bus, 37
VxWorks, 38

W

Wave number, 98
Wiggling effect, 92
    Compensation, 96–100

I want morebooks!

Buy your books fast and straightforward online - at one of the world's fastest growing online book stores! Environmentally sound due to Print-on-Demand technologies.

Buy your books online at
www.get-morebooks.com

Kaufen Sie Ihre Bücher schnell und unkompliziert online – auf einer der am schnellsten wachsenden Buchhandelsplattformen weltweit!
Dank Print-On-Demand umwelt- und ressourcenschonend produziert.

Bücher schneller online kaufen
www.morebooks.de

SIA OmniScriptum Publishing
Brivibas gatve 1 97
LV-103 9 Riga, Latvia
Telefax: +371 68620455

info@omniscriptum.com
www.omniscriptum.com

Printed by Books on Demand GmbH, Norderstedt / Germany